THE DUKE'S CUT
THE BRIDGEWATER CANAL

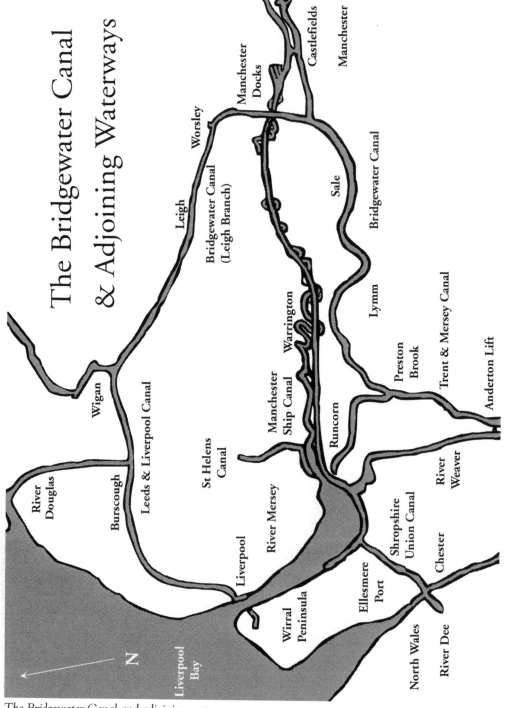

The Bridgewater Canal
& Adjoining Waterways

The Bridgewater Canal and adjoining waterways.

THE DUKE'S CUT
THE BRIDGEWATER CANAL

Cyril J. Wood

The
History
Press

The Manchester Ship Canal Coat of Arms.

First published 2002
This edition first published 2009
Reprinted 2013

The History Press
The Mill, Brimscombe Port
Stroud, Gloucestershire, GL5 2QG
www.thehistorypress.co.uk

British Library Cataloguing in Publication Data.
A catalogue record for this book is available from the British Library.

ISBN 978 0 7524 5111 4

Typesetting and origination by The History Press
Printed in Great Britain

Contents

Acknowledgements

I would like to thank my wife, Angie, for putting up with me whilst working on this project and for her painstaking help with the proof-reading; Mike Webb in the Bridgewater Department of Peel Holdings for his invaluable assistance, corrections, generosity and for allowing access to the Bridgewater Canal Photographic Archives; Margaret, Brian and Nigel Hamilton of Thorn Marine, Stockton Heath for lending photographs for scanning; and the many friends and fellow canal cruisers too numerous to mention who have also loaned photographs for scanning, and provided additional details and information. All contemporary photographs were taken by myself unless otherwise accredited.

Introduction

I have been interested in canals and inland waterways since a child when my parents hired canal cruisers on the Shropshire Union and Llangollen Canals before having a boat of their own built. I followed their example, bought my first boat in 1983 and moored it on the Shropshire Union Canal at Beeston Iron Lock before moving to the Bridgewater Canal. I was already aware of the historical impact that this canal had made on inland navigation in this country and the more I cruised the Bridgewater Canal, the more I appreciated its significance, diversity and grandeur.

I started to photograph the canal and gave audio/visual presentations on it to various societies. This graduated into mapping the whole canal and writing about it with a view to having the finished work published. It would be very pretentious of me to say that this is the definitive publication on the canal as each book has its own individuality and focuses on different aspects of the canal. I have tried to produce a book that concentrates on the 'mechanics' of the canal's history and geography concisely, without the encumbrance of facts that the reader usually skips.

I hope that you, the reader, gain as much enjoyment out of reading this book as I have had producing it and that you find it a readable, informative and entertaining piece of work that relates the canal's history, describes its route, gives invaluable information to those wishing to use it and documents the canal with photographs of features that are familiar or have disappeared and of places that have changed beyond recognition.

I apologise in advance for any mistakes or inaccuracies that have may have crept into the text, maps or photographs in this book.

Cyril J. Wood
February 2002

Chronology

84AD	Fosse constructed at Castlefield by Romans
1737	Scroop Egerton commissioned Thomas Steers to investigate making Worsley Brook and mine soughs navigable
1745	Improvements made to River Douglas
1754	Survey to make Sankey Brook navigable by Henry Berry and John Gilbert first employed by Francis Egerton as land agent
1757	Saint Helens Canal partially open
1757	John Gilbert appointed mine manager at Worsley
1757	Francis Egerton takes up residence at Worsley
1758	James Brindley introduced to Francis Egerton by John Gilbert and visited Worsley. Also, work commenced on the Bridgewater Canal at Worsley on the Duke's land
1758	James Brindley completes survey for Bridgewater Canal
1759	First Bridgewater Canal Act submitted to Parliament
1-7-1759	Construction commenced on Bridgewater Canal outside Worsley
11-1759	Terminus changed from Salford Quay to Dolefield and the necessary Act of Parliament passed
17-7-1761	Opening of Barton Aqueduct
1763	Terminus changed yet again from Dolefield to Castlefield
1763	Connection made to Mersey and Irwell Navigation at Cornbrook (the Gut)
8-1761	James Brindley surveyed the line from Stretford to Runcorn
3-1762	Act of Parliament passed allowing construction of line from Stretford to Runcorn
7-1765	Canal constructed to Castlefield
1766	Act of Parliament passed allowing construction of the Trent and Mersey Canal including deviation of the Bridgewater Canal allowing a junction at Preston Brook
1767	Passenger carrying commenced on the Bridgewater Canal
1768	Construction of the new line reached Lymm
1769	Construction at Norton Priory near Runcorn held up by Sir Richard Brook
27-9-1772	James Brindley died at the age of 56 caused by a chill aggravating his diabetes
1772	Construction of the locks to the River Mersey at Runcorn completed
1772	Bridgewater Canal and Trent and Mersey Canals connected at Preston Brook
1775	The disagreement with Sir Richard Brook resolved by intervention by Parliament, allowing completion of the canal to Runcorn
1-1776	Canal completed at Runcorn
21-3-1776	Canal opened to through traffic
1791	Act of Parliament passed and construction of the Leeds and Liverpool Canal commenced
1794	Act of Parliament allowing construction of the Rochdale Canal
1799	Old Hollins Ferry branch of the Bridgewater Canal at Worsley extended to Pennington
1800	Rochdale Canal reaches Castlefield
1800	Ashton Canal opened
1801	Duke observes experimental tug "Charlotte Dundas"
8-3-1803	Francis Egerton dies, George Gower inherits the income from the Bridgewater Estate and a trust set up to look after the Bridgewater Canal
1804	Rochdale Canal completed
1819	Act of Parliament allowing extension of the Leeds and Liverpool Canal to be extended to join the Bridgewater Canal at Leigh
1821	Leeds and Liverpool and Bridgewater canals joined at Leigh
1822	Proposal for the Liverpool to Manchester Railway and Barton Aqueduct damaged by floods necessitating rebuilding

1823	Extension of the Bridgewater Canal from Sale to Stockport proposed and objected to by Ashton and Peak Forest Canals
1824	Rebuilding work on Barton Aqueduct completed
1825	Proposal for ship canal from Runcorn to West Kirby promoted and subsequently dropped mainly due to cost
1825	Act of Parliament for Liverpool to Manchester Railway submitted and subsequently thrown out
1826	Liverpool to Manchester Railway Act of Parliament passed
1827	New line of locks constructed at Runcorn
1827	New warehousing and facilities built at Castlefield and Preston Brook
1830	Liverpool to Manchester Railway completed
1833	George Gower dies and Bridgewater Estates income inherited by Lord Francis Leveson-Gower, on the understanding that he changed his name to Lord Francis Egerton.
1837	Lord Francis Egerton and his wife Harriet take up residence at Worsley
1838	Hulme Lock branch built superseding the previous connection to the River Irwell… The Gut
1838	Experimental steam tugs introduced into the Bridgewater Canal
1838	Proposal for extension from Altrincham to Middlewich
1844	Lord Francis Egerton purchased Mersey and Irwell Navigation shares
17-1-1746	Act of Parliament allowing the Bridgewater Estate to take over the shares of the Mersey and Irwell Navigation from Lord Egerton
1851	Queen Victoria travels along the Bridgewater Canal from Patricroft Station to Worsley
1865	Steam tugs introduced for towing through Preston Brook Tunnel
1869	Prince of Wales travels on the Bridgewater Canal from Worsley to Royal Agricultural Show at Trafford
1872	Bridgewater Navigation Co Ltd formed to raise capital for investment into Mersey and Irwell Navigation and Bridgewater Canal
1877	Hamilton Fulton proposes a ship canal to Manchester
1882	Manchester Ship Canal proposed by Daniel Adamson
1885	Bridgewater Navigation Co Ltd purchased by newly formed Manchester Ship Canal Co
11-11-1887	Construction commenced on Manchester Ship Canal
1888	Worsley mines cease coal mining but remain open for drainage, ventilation and access
1890	Sprinch's Boatyard opens at Runcorn
1-1-1894	Manchester Ship Canal opens
1935	Sprinch's Boat Yard burnt down
1839	Old Line of Locks at Runcorn closed
1948	Old Line of Locks in-filled
1952	Pleasure craft allowed on the Bridgewater Canal
1952	Bridgewater Motor Boat Club formed at former Sprinch's Boat Yard site at Runcorn
1955	Lymm Cruising Club formed
1958	Worsley Cruising Club formed
1959	Sale Cruising Club formed
22-3-1964	Watchouse Cruising Club formed
1966	New Line of Locks in-filled at Runcorn
1971	Bollin Aqueduct breach
1974	Commercial traffic ceases on Bridgewater Canal
1974	Preston Brook Marina opens
11-1981	Preston Brook Tunnel collapses
4-1984	Preston Brook Tunnel re-opens
1984	Bridgewater Estates purchased by Peel Holdings
1985	Preston Brook Boat Owners Association formed
1987	Development at Castlefield commences
1988	IWA National Rally held at Castlefield
1995	New Pomona Lock replaces Hulme Lock as connection with the River Irwell, Manchester and Salford Docks and the Manchester Ship Canal
2005	IWA National Waterways Festival at Preston Brook
2-2006	New marina in Stretford completed

One

The History of the Bridgewater Canal

The first inland navigations in England can be attributed to the Romans, who constructed navigable cuts, known as 'fosses' or 'dykes', in some of our rivers to bypass navigational hazards. Three of the better known of these cuts are the Caer Dyke, the Fosse Dyke and the Itchen Dyke.

The Carr (or Caer) Dyke ran from the River Witham at Lincoln to Peterborough and the Fosse Dyke also ran from the River Witham at Lincoln but connected the town with the River Trent. The Itchen Dyke ran from Winchester to the sea. No doubt these artificial waterways were monumental in the development of the Lincoln area.

Another fosse that is often overlooked in the history of our inland waterways was built in the Castlefield area of Manchester to connect the Rivers Irwell and Irk. This particular fosse was built around AD 84 and, in building this waterway, the Romans laid the foundation stone for a series of waterways; the development of such a system was to have a profound effect on transport in this country at the start of the Industrial Revolution and beyond. Unfortunately, no remains of the Castlefield Fosse can be found.

Over the successive centuries there have been a few other attempts to produce workable navigations. The erroneously named Exeter Ship Canal, constructed in 1566 by John Trew, ran alongside the River Exe, but could only accommodate barges. It was originally constructed to bypass a section of the river, notorious for shoals and scours, connecting Exeter to the sea. This navigation featured the first pound locks (as opposed to flash locks) in England. Pound locks are often attributed to Leonardo Da Vinci but there have been locks of this type in Holland since the fourteenth century, long before Da Vinci's birth.

In 1754 a survey of the Sankey Brook at St Helens was carried out by Henry Berry, a local engineer. Work soon started on the brook to make it navigable from St Helens to the River Mersey, and by 1757 the St Helens (or Sankey) canal was partly open.

Whilst being called a canal, this and other early works utilised an existing watercourse, whether it was a brook, stream or river, and is thus not a canal in the true sense of the word but a 'navigation'. The general consensus of opinion in inland waterway circles is that the first

Francis Egerton, Third Duke of Bridgewater.

canal built was the St Helens Canal, but this is open to speculation. The definition of a canal is a waterway constructed independent of any existing watercourse except for the water supply. As the St Helens Canal is mostly the canalised Sankey Brook, it is classed as a navigation. The first true canal in Britain built independent of a watercourse was the Bridgewater Canal or, as it is more affectionately known, 'The Duke's Cut'.

The history of inland waterways navigation in the area surrounding the Bridgewater Canal is complex and many schemes overlapped each other both in the historical and geographical context. For instance, the plans to make the River Douglas navigable postponed the plans for the Bridgewater Canal but figured in the development of the Leeds & Liverpool Canal. The River Irwell and subsequently the Mersey & Irwell Navigation were, one hundred and fifty years later, to be absorbed into one of the most ambitious canal projects in England – the Manchester Ship Canal.

Scroop Egerton, the first Duke of Bridgewater, owned mine workings at Worsley near Manchester, and required a reliable means of transporting his coal that was not dependent on using pack horses and carts via the notoriously unreliable roads. In 1737, the Duke commissioned Thomas Steers to investigate the practicality of making the mine's drainage soughs and Worsley Brook navigable as far as the River Irwell. Steers had considerable experience in civil engineering; he was responsible for Liverpool's first dock and a survey to make the River Irwell navigable to Manchester. The Duke's plan was dropped when improvements were made to the nearby River Douglas and he died in 1745.

Francis Egerton was the sixth son of Scroop and became the third Duke of Bridgewater at the age of eight when his father died. He was a sickly child and suffered an unhappy childhood, brought up by his mother and stepfather who showed little interest in the young Francis, and punctuated by episodes of tuberculosis. He was educated at Eton until the age of sixteen when Robert Wood was appointed the young Duke's tutor. Wood agreed to take the young Duke on a 'Grand Tour' where he would be exposed to the sights and different cultures of a world outside that of England. Francis showed an interest in the transportation system of France and visited the Grand Languedoc Canal (later known as the Canal du Midi) which connects the Atlantic Ocean at Bordeaux with Sète on the shores of the Mediterranean.

The underground mine entrances at Worsley Delph.

11

Brindley's original Barton Aqueduct.

Once the Grand Tour was over, the young Duke returned to England and was absorbed into London's social life. It was here that he met the widowed Lady Elizabeth Hamilton (one of the Gunning sisters, well known in London's social circles) with whom he had a love affair and was to become engaged to. The Duke dissolved the engagement due to a scandal in the Gunning family concerning Elizabeth's sister, a decision that caused him much heartache.

Dismayed with London life, the Duke decided to dedicate his life to commerce. In 1757 he left London and headed for his mines at Worsley near Manchester where he took up residence in the family estate at Worsley Hall. The relationship with his mother and his affair with Lady Hamilton had scarred him emotionally and he became a misogynist; hating women to such an extent that he would not even allow female servants at the Hall.

After settling in to his new life, he met with John Gilbert, the mine's agent and engineer, to discuss problems associated with the mines. It transpired that the two main problems were mine drainage and transportation of the coal. At that time, the coal was carried by cart or packhorse to the River Irwell where exorbitant charges were made to transport the coal to Manchester, its main market.

Egerton had seen the success of the Grand Languedoc Canal and other continental waterways on his Grand Tour and was, no doubt, aware of the nearby fledgling St Helens Canal. Together he and Gilbert resurrected Scroop Egerton's idea for a canal; they expanded it to incorporate an elaborate drainage system for the mines and began to survey the route. The proposed canal would not only solve the transportation problem and alleviate that of the mine's drainage, it would bring down the price of Worsley coal in Manchester, thus making it more competitive and an affordable commodity for the less affluent members of society.

The following year, Egerton was introduced to a millwright called James Brindley by Gilbert. Brindley travelled to Worsley and stayed as the Duke's guest for six days discussing mine drainage and the proposed canal. Five years earlier, Brindley had been involved in extensive mine drainage works at the not too distant Wet Earth Colliery near Clifton (adjacent to the currently disused Manchester, Bolton & Bury Canal). No doubt, the success of this scheme helped to make up the Duke's mind when he decided to engage Brindley to complete the survey for the canal already started by Gilbert.

12

Worsley Delph with the two mine entrances obscured by foliage. Note the sunken Starvationer on the far left.

Close-up of the sunken Starvationer showing the exposed ribs that give the boat its name.

Brindley soon completed the survey and in 1759 an Act of Parliament was passed, enabling the canal to be built. The proposed route was to run from the mines at Worsley to the River Irwell at Salford Quay with a branch to Hollins Ferry, also on the Irwell, 9.6km (6 miles) below Barton Bridge. On 1 July 1759 work on the canal commenced, but in November of that year the Salford Quay terminus was dropped in favour of Dolefield and an additional Act of Parliament obtained.

There is an interesting offshoot to the early history of the Bridgewater Canal. Francis Egerton's brother-in-law was Lord Gower, who owned coalmines and limestone quarries in the rapidly expanding industrial area of Shropshire. His land agent was John Gilbert's brother, Thomas Gilbert. It is obvious that communication had taken place between the two families and lead to the foundation of the canal system in Shropshire, which resulted in the area being christened 'The Cradle of the Industrial Revolution'.

Around this time, another canal was being proposed. This was the Grand Trunk or, as it was later known, the Trent & Mersey Canal. This canal was the brainchild of Josiah Wedgwood and would be used to bring clay to his potteries and transport the finished pottery, while also serving the Cheshire salt industry. It is interesting to note that two of the main promoters of the canal were none other than Lord Gower and Thomas Gilbert. Wedgwood had also secured the services of James Brindley to survey and build the canal. Brindley saw the Grand Trunk as the foundation stone for a system of canals connecting the Rivers Trent, Mersey, Severn and Thames, with other canals branching off to serve various towns and cities. Hence the name 'Grand Trunk', like a tree with branches spreading out to various parts of the country.

The mines at Worsley were constantly being expanded downwards and outwards in order to reach the seams of coal that ran throughout the area. Trips were even arranged for the adventurous to view the workings first hand for a nominal fee. Meanwhile, construction on the Bridgewater Canal had been progressing from both ends. Several engineering hurdles had been overcome and the canal had reached a major obstacle, the River Irwell. Initially, a flight of locks were planned to lower the canal to the river with another flight to raise it up on the

Left-hand mine entrance at Worsley.

14

other side. This would have used too much of the canal's water resources so Brindley planned to bridge the river using a masonry aqueduct (the stone being waste obtained from the Worsley Mines) lined with puddled clay (wet clay kneaded like dough) to make it waterproof.

When the Act for the aqueduct was proposed, MPs reading the proposal called Brindley into their chambers at the House of Commons to demonstrate how he expected to make the bridge waterproof. Brindley was semi-literate and often resorted to drawing diagrams with whatever material was at hand. In addition to drawing on the polished floor of the Parliamentary Chambers, he demonstrated how the aqueduct would work with the aid of a large cheese, which he carved to represent the aqueduct, filling it with water when the model was completed.

Brindley demonstrated the theory of clay puddling by bringing buckets of water and wet clay into the chambers. He gave a practical demonstration of how clay could be made waterproof by puddling and could then be applied to the masonry to make it, in turn, waterproof. Needless to say, the Act was passed and on 17 July 1761 water was admitted to the completed Barton Aqueduct, which opened the Bridgewater Canal from Worsley to Stretford on the outskirts of Manchester.

Barton Aqueduct was a three arched masonry structure, 183m (600ft) long, 11m (36ft) wide and 12m (39ft) high. Scepticism was rife prior to its opening. It was given the nickname 'Castle in the Air' and many people thought that it would surely collapse when water was admitted. Needless to say, Brindley proved the sceptics wrong although there was a problem when one of the arches started to bulge. This necessitated the drainage of the aqueduct after the opening ceremonies but was soon rectified and the canal was open to through traffic. When posed with a problem, Brindley would often retire to his bed and ponder the best course of action. On this particular occasion, when he rose from his deliberations, John Gilbert had already solved the problem for him, completed the remedial work, and refilled the aqueduct. Brindley's comments on this action are not documented.

The building of the remainder of the proposed route was progressing well and in 1763 the terminus was changed yet again making Castlefield near Deansgate the eventual terminus. Also in this year, a connection at Cornbrook was made with the Mersey & Irwell Navigation.

Right-hand mine entrance at Worsley.

Waterloo Bridge, Runcorn, viewed from the River Mersey side. Note the drydock beneath the left-hand arch containing a Bridgewater Tug for maintenance.

This connection was known as 'The Gut', but could only be used when there was sufficient water in the river to overcome the silting and low river water levels. Castlefield was reached in August 1765 and, in doing so, marked the completion of the first canal in England to be constructed independent of a watercourse.

During the period of construction (and afterwards) the Duke had amassed a considerable debt. He borrowed money from many people to finance the building of the canal. When not in residence at Worsley, he was constantly visiting possible contributors in an effort to raise extra funding. It was to be many years before his financial labours bore fruit, the debts were satisfied and the canal returned a profit.

Meanwhile, Brindley had not been idle. In addition to his work on the Trent & Mersey section of his Grand Trunk scheme, he had started work on the Staffordshire & Worcestershire Canal, which would connect the Trent & Mersey Canal at Great Heywood Junction near Cannock Chase to the River Severn at Stourport. In September 1761 he started the survey on another venture for the Duke of Bridgewater. This was a canal running from a junction with his existing canal at Stretford to join the River Mersey at Runcorn. The line of this canal would prevent the Mersey & Irwell Navigation's monopoly in this area. The locks at Runcorn would also provide a connection to the proposed Duke's Dock at Liverpool, which was not to be completed until 1774. At this time the decision was made to abandon the branch at Hollins Ferry even though two miles of it had already been cut from Worsley.

In March 1762 an Act of Parliament was passed for the Runcorn Line and work started immediately. Another Act of Parliament, which concerned the Bridgewater Canal directly, was the Trent & Mersey Canal Act of 1766. In addition to permitting construction of the canal, it made a change to the Bridgewater's original line, allowing the two canals to link up at Preston Brook near Runcorn.

In the following year, passenger traffic commenced on the Bridgewater Canal in specially constructed 'Packet' boats. The boats were of lighter construction than conventional craft to facilitate greater speed and even possessed a sharp knife on the bows to cut the towropes of any

Waterloo Bridge viewed from the Preston Brook side. The dry dock occupies the centre arch.

Preston Brook Tunnel entrance showing the now demolished house that used to be on top of the entrance.

other craft that dared to get in the way. This service proved to be very popular and as the canal extended in length so was the service.

After various challenges, such as the aqueduct over the River Mersey and the crossing of the marshy Sale Moor, Lymm was reached in 1768. However, in the following year, construction at the Runcorn end came to an abrupt standstill when Sir Richard Brook, a local landowner, objected to the canal's presence on his estate at Norton Priory. This held up construction for a few years. Meanwhile, at the other end of the canal, Stockton Heath had been reached. Whilst overseeing construction of the Runcorn end of the canal, the Duke and Brindley both stayed at the imposing Bridgewater House. This was a residence that the Duke had built at Runcorn, situated adjacent to the flight of locks connecting the canal to the River Mersey.

Brindley was a diabetic and his health was suffering. He was continually commuting between his other canal projects at Oxford, Coventry, Chester, Calder and Hebble, among others. His mode of transport was by horseback, spending nights at inns and travelling in all weathers. He caught a chill, which added to his diabetes, and he died on 27 September 1772 at the age of fifty-six.

In 1772 the flight of ten locks leading down to the River Mersey at Runcorn was completed. However, they could not be used until the disagreement with Sir Richard Brook had been resolved. This disagreement lasted until 1775 when Parliament intervened and a settlement was reached. The impact of this settlement on the canal still remains with us to this day. This is the deviation of the canal's route in the shape of an 'S' bend around Sir Richard Brook's land and the changing sides of the towpath between Old Astmoor Bridge and Norton Bridge. The only other place on the Bridgewater Canal where the towpath changes sides is in Manchester between Old Trafford and Castlefield.

Another memorable event in 1772 was the linking of the Bridgewater with the Trent & Mersey Canal. The actual place where the two canals meet is 10m (11yd) inside Preston Brook Tunnel. The spot is marked by a Trent & Mersey Canal milepost on the horse path that goes over the top of the tunnel. In the absence of a towpath, boats were originally 'legged' through the tunnel until a steam tug service was introduced. Today, the entire length of the tunnel

Preston Brook Bridge prior to rebuilding, which now carries the A56.

Aerial view of Runcorn showing the Old and New Lines of Locks. Bridgewater House is in the lower left and the Manchester Ship Canal is in the bottom of the photograph, complete with sailing craft.

Bridgewater House, Runcorn.

The Linotype Works at Altrincham.

comes under the jurisdiction of British Waterways with the Bridgewater Canal commencing at the stopboards at the northern portal.

In January 1776, the final mile through Norton Priory was cut and the canal opened for through traffic on 21 March. It was a pity that James Brindley was not alive to see his canal completed.

The latter part of the eighteenth century saw Acts of Parliament for two canals that would eventually link with the northern end of the Bridgewater Canal. They were the Leeds & Liverpool Canal in 1791 and the Rochdale Canal in 1794. It seems coincidental that two of the three cross-Pennine waterways should be directly connected to the Bridgewater Canal, and the third, the Huddersfield Narrow Canal, connected via the Rochdale and Ashton Canals. The Leeds & Liverpool Canal was started in 1791 but it was not until 1799 that part of the abandoned 1759 Hollins Ferry extension of the Bridgewater Canal was extended to make an end-on junction with the Leeds & Liverpool at Leigh. The part of the Hollins Ferry extension not used in the Leigh Branch was to be used as a dump for dredgings, the remains of which can still be seen today just outside Worsley.

The year 1795 must be mentioned for a much sadder reason: the death of John Gilbert, agent and engineer for the Bridgewater Mines and Canal since 1757 and who we must give some credit as one of the major inspirations behind the Bridgewater Canal, along with Francis Egerton and James Brindley.

From the very beginning of the canal's history, horses had been the prime motive force, but in 1796 an experimental steamboat was commissioned by the Duke in an attempt to speed up transport along the canal. By 1799 the craft was completed and sailed along the canal for the first time. The boatmen along the canal looked upon the experimental tug with scorn, fearing that this technological wonder would make their jobs redundant. They had little to worry about as the stern-mounted paddle wheel of the craft produced a large wash that would have eroded the canal banks over a period of time. The steamboat was withdrawn and broken up although its engine was retained to power a water pump. Although not a success, the craft was to herald the change to steam propulsion that took place over the next forty years.

By 1799, the Leigh Branch had reached Pennington and in 1800 the Rochdale and Ashton Canals were opened to Castlefield. The Rochdale Canal, however, was not completed and did not cross the Pennines until 1804, the year after the Duke's death.

Remains of Brindley's Cloverleaf Weir at Potato Wharf, Castlefield.

The third Duke was well liked and treated his workers with kindness and (on the whole) respect. Even so, he was prone to eccentric behaviour. He swore profusely, seldom washed and wore the same clothes, day in and day out. Concerned about his employees' timekeeping, he arranged for the clock in a turret of the works yard at Worsley (later moved to Worsley Church) to strike thirteen at one o' clock signalling the end of lunch hour. On one occasion he met a miner who was late for work and questioned him. The worker informed him that his wife had given birth to twins during the night to which the Duke said, 'Ah well, we must accept what the Lord sends us', and the miner replied, 'Aye, but I notice he sends all the babies to our house and all the brass to yours!' This reply may have pricked the Duke's conscience as well as making him laugh. He consequently gave the miner a guinea. As wages were paid monthly, he arranged for employees' wives to buy their groceries on account at a local shop, the bill being deducted from the workers' salaries and the balance given in cash, thus preventing the wages from being squandered in local hostelries. The Duke also formed a 'sick club' to which his workers could contribute.

The Duke saw a demonstration of the *Charlotte Dundas* tug in 1801 and was so impressed that he ordered eight similar craft to be built. Unfortunately the order was cancelled in 1803 after the Duke's death and before any of the craft had been delivered.

The Duke died on 8 March that year. His will laid down that the main beneficiary of the canal would be his nephew, Earl George Gower, later to become the Duke of Sutherland. It also stated that a Board of Trustees was to be formed to look after the interests of the Duke's estate. The trustees were prominent national figures, many of whom had a vested financial interest in the maintenance of the Bridgewater Canal overseen by George Gower. Later, the trustees of the canal were to be local councils, whose provinces the canal passed through, and this remained in place until 1903. On the death of George Gower in 1833, the profits from the Bridgewater Estate went to his second son, Lord Francis Leveson-Gower, on the understanding that he changed his name to Lord Francis Egerton.

In 1819 the Leeds & Liverpool Canal Act to build a branch that joined the Bridgewater Canal was passed. The two canals were to meet at a head-on junction at Leigh, which was completed two years later. In the following year, 1822, a proposal for a railway from Manchester

Warehouses at Castlefield.

The entrance to Hulme Lock, Manchester showing a Bridgewater Barge and a swing bridge on the right.

to Liverpool was put forward. From the outset, the Bridgewater Canal opposed its development, seeing the far-reaching repercussions it would have on canals.

In an effort to make its route more comprehensive, two extension plans were drawn up. The first plan in 1823 proposed an extension from Sale to Stockport but was thwarted due to oppositions from the boards of the Ashton and Peak Forest Canals. The second plan, two years later, was far more ambitious. It was to be a canal, possibly of ship canal dimensions, to link Runcorn on the Mersey with West Kirby on the River Dee coast of the Wirral Peninsula. This plan was thrown out for many reasons, the major one being cost. It is possible that the latter plan was to have been carried out by Thomas Telford who, coincidentally, later carried out a survey to construct a ship canal from Wallasey Pool to West Kirby, which would bypass the navigational hazards of the River Mersey Estuary. Telford is reported to have said about Liverpool, 'Look, they've built the docks on the wrong side of the river', due to Wallasey Pool, the location of Wallasey and Birkenhead Docks, being a natural harbour. Had his plan been successful, Liverpool would undoubtedly not have had the successes that it ultimately enjoyed.

The Liverpool & Manchester Railway Company went to Parliament with their Act in 1825, only to have it opposed by a joint objection from the boards of the Bridgewater Canal and the Mersey & Irwell Navigation. However, the respite was only a temporary one and the Act was successfully passed the following year.

Sensing that a battle was at hand, the Bridgewater Canal started a regime of expansion. A new line of locks at Runcorn was built next to the existing ones in 1827. Along with this, new warehousing facilities at Runcorn, Preston Brook, Castlefield and many other locations were built. The aim behind this regime was to speed up cargo handling and hopefully be more competitive when the Liverpool & Manchester Railway was completed in 1830.

In 1833, amidst the railway threat and expansion of the canal's cargo facilities, George Gower died. His son then changed his name to Lord Francis Egerton, First Earl of Ellesmere, and took up residence at Worsley with his wife Harriet in 1837. Lord Egerton was as concerned about the welfare of the workers as his namesake. His wife shared his concern and encouraged her husband to abolish the use of child labour in the mines.

The Duke of Bridgewater in later years.

Over the years, Castlefield had been modified considerably. The main water supply to this end of the canal was the River Medlock. At the end of the basins near Deansgate, the Medlock plunged into a syphon; the tunnel took the river beneath the basins to emerge beyond Potato Wharf. To unload, the boats were steered into a tunnel hewn into the rock face. The river drove a waterwheel that powered a winch used to raise goods from the boats below up to street level. When the Rochdale Canal was built, its bed cut across the path of the tunnel, which eventually fell into disuse. The entrance to the tunnel has been rebuilt, as has the waterwheel, and its operation can be demonstrated by guides showing visitors around the area. Part of the other end of the tunnel can be seen at Pioneer Wharf, adjacent to Deansgate, as an arch two or three feet above the water level of the Rochdale Canal. Whilst the Manchester Ship Canal was being constructed in the 1890s, the syphon method was again utilised to convey rivers, such as the River Gowy at Stanlow, beneath the canal.

Another structure that shrunk in size as the wharves expanded was Brindley's original 'cloverleaf' weir (so called because of its shape). This weir returned excess water from the canal to the River Medlock as it emerged from the syphon. In 1838, most of the weir was removed for two reasons: it frequently blocked and the space was needed for additional wharves.

In 1838 the Hulme Lock Branch was built. This branch connected the Bridgewater Canal to the River Irwell adjacent to where it joins the River Medlock. The Hulme Lock Branch superseded the previous connection with the Mersey & Irwell at Cornbrook known as 'The Gut'. The same year also saw the introduction of experimental steam tugs on the canal and a proposal for a branch to run sixteen miles from Altrincham to Middlewich. It would cross the Trent & Mersey Canal to join the Middlewich branch of the Chester & Ellesmere Canal (later known as the Shropshire Union Canal). The board of the Chester & Ellesmere Canal encouraged the Bridgewater Trustees in this proposal but it came to nothing (as could be expected) due to opposition from the Trent & Mersey, Macclesfield, Peak Forest and Ashton Canals boards.

Despite the modernisation of cargo handling facilities and a toll war with the Mersey & Irwell Navigation, the railways were having a noticeable effect on the tonnage carried along the

The location of the syphon where the River Medlock plunges into a tunnel.

Close-up of the weir and the start of the syphon tunnel.

Bridgewater Canal. One logical solution to the problem would be to control trade on the Mersey & Irwell Navigation. They were having the same difficulties as the Bridgewater Canal, as the drop in dividends paid to their shareholders could testify.

Consequently, Parliament was applied to for an Act enabling the Bridgewater to purchase the shares of the Mersey & Irwell. This Act was passed and the transfer of shares took place on 17 January 1846. The sum paid for the navigation was £550,000.

The Bridgewater Canal, being a commercial waterway, was navigated twenty-four hours a day, and did not generally allow pleasure craft on its waters. However, there are always exceptions to the rule, The first took place in 1851 when Queen Victoria took a trip along the canal from Patricroft to Worsley where she was to be entertained at Worsley Old Hall. The Queen refused to cross Barton Aqueduct, saying that it would be more like flying than sailing. The canal's normal colour at Worsley is ochre, caused by the iron deposits from the water draining out of the mines but the canal was dyed royal blue in honour of the Queen's visit.

The second exception was in 1869 when the Prince of Wales took a trip from Worsley to Trafford for the opening of the Royal Agricultural Show. On both these occasions the craft used was the *VIP Barge*, later to be used as an inspection launch. The boat was permanently kept at Worsley in the boathouse adjacent to the dry docks. It was later converted to diesel power and was in use until 1948 when it was broken up. In later years, the Duke encouraged his rich landowner friends to have 'gondolas', as he called them, built for use on the canal for leisure purposes.

Whilst the acquisition of the Mersey & Irwell Navigation undoubtedly helped to boost the canal's financial position in the face of the railways, it still needed to be competitive. By 1860, the Mersey & Irwell was so silted up that it was impossible for all but the shallowest drafted craft to reach Manchester. Consequently, in 1872, a new company was formed to inject capital necessary for the dredging of the Mersey & Irwell and increased warehousing, as well as new cargo handling equipment on the Bridgewater Canal. This also included the purchase of the steam tugs after their successful trials had been completed. The new company was called the Bridgewater Navigation

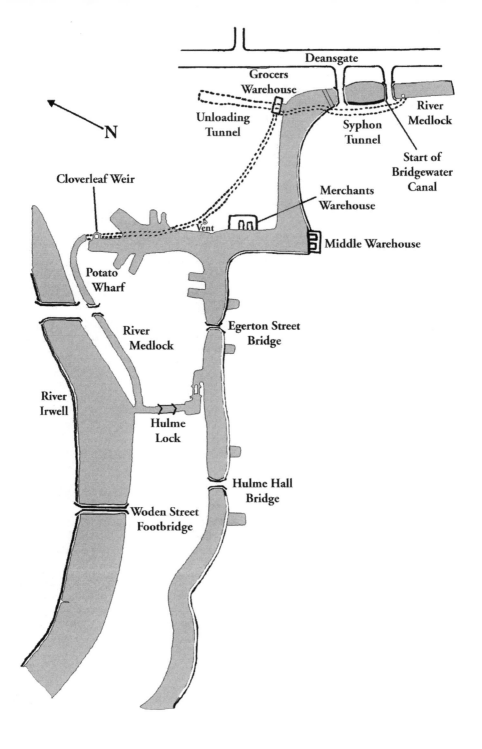

Castlefield area prior to the construction of the Rochdale Canal.

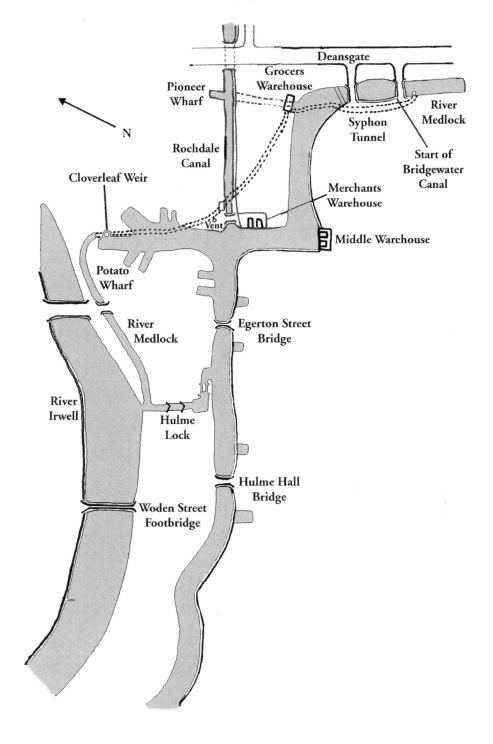

Castlefield area after the construction of the Rochdale Canal.

Company Limited. It is ironic that many of the shareholders also held shares in railway companies and that the body collecting shares for the trustees was a collection of railway companies.

The tugs that were purchased were the famous 'Little Packets'. These were steam-powered vessels, 18.3m (60ft) in length, powered by a single cylinder horizontal engine driving a 1m (3ft) propeller. Initially, five tugs were ordered and they were so successful that by 1881 their numbers had risen to twenty-six. These vessels were in regular use on the Bridgewater and adjacent waterways (hence the 18.3m/60ft length) until the 1920s when some were converted to diesel power and the remainder were either sold or scrapped. Those converted to diesel power soldiered on until the 1950s when more modern replacements were purchased. For many years, the hull of one of these tugs could be seen out of the water at one of the 'wides' (lakes connected to the canal caused by salt mining subsidence) on the Trent & Mersey Canal near Middlewich before being bought for restoration. The man behind the introduction of the tugs was the new General Manager and Engineer of the Bridgewater Navigation Company Limited, Edward Leader Williams.

Prior to 1865, craft were legged through Preston Brook Tunnel. This involved a plank laid across the bows of a narrow boat on which two men would lay on their backs and propel the boat through the tunnel by 'walking' along the tunnel roof. Gangs of men were employed solely for this task.

In the early part of 1865, steam tugs similar to the type used for towing on the canal were introduced for tunnel passage duties. Not long after their introduction, a tug driver and his stoker were overcome by smoke caused by the lack of ventilation inside the tunnel. Consequently, Mr Forbes, the Trent & Mersey Canal's Resident Engineer, ordered four ventilation shafts to be sunk to ease the situation. Whilst the shafts were being sunk, traffic (including the steam tugs)

The Rochdale Canal drained for maintenance, Pioneer Wharf is on the left.

The end of Pioneer Wharf showing the remains of Brindley's unloading tunnel, which was bisected when the Rochdale Canal was built.

continued to pass through the tunnel. In May 1865, two maintenance workmen constructing the vents, hitched a work boat onto the end of a train of boats being towed through the tunnel by a steam tug. They too were overcome by the smoke; one man fell into the tunnel and subsequently drowned. At an inquest into the workman's death, the coroner ordered that the tugs must not be used until the ventilation shafts were completed.

Another type of familiar craft on the Bridgewater were the passenger or 'Fly' boats that plied the length of the canal daily. One such craft was the *Duchess Countess* built in 1871. The last resting place of this craft was as a hen house at Welsh Frankton on the banks of the Llangollen Canal until it was broken up in the early 1960s.

The Bridgewater Navigation Company Limited was a fairly short-lived company. In 1885, the Bridgewater and Mersey & Irwell Navigations were bought by the newly formed Manchester Ship Canal Company. The Manchester Ship Canal was proposed by Daniel Adamson, a Manchester businessman, in 1882 and was to be a waterway capable of conveying ocean-going ships from Eastham on the River Mersey to a proposed dock complex close to the centre of Manchester. It would follow the same route as the Mersey & Irwell Navigation from Runcorn to Manchester with several new cuts and extensions, plus a completely new section from Eastham on the Wirral to Runcorn, effectively hugging the banks of the River Mersey Estuary. The plans were very similar to an earlier scheme proposed by Hamilton Fulton in 1877 and the plan was vigorously opposed by many groups, including the City of Liverpool who were afraid that the competition from the Ship Canal would adversely effect the city's livelihood. After a long, protracted battle, the Act of Parliament was passed on the third reading and Lord Egerton cut the first sod of earth on 11 November 1887.

The construction of the Ship Canal directly affected the Bridgewater Canal in two ways. The first effect was at Runcorn where the River Mersey could only be accessed by crossing the Ship Canal to Bridgewater Lock or by sailing down its length to Eastham. The second change was at the famous Barton Aqueduct, which would have to be demolished due to the limited headroom of Brindley's original structure.

Projected schemes for the aqueduct's replacement included locks to lower craft to the level of the Ship Canal and up the opposite side (as in the Bridgewater Canal's original proposal) and a vertical lift similar to that at Anderton connecting the Trent & Mersey Canal with the River Weaver. The

A steam-powered Bridgewater Tug towing a train of barges in Walton Cutting. The bridge to Walton Hall is just visible in the background.

The derelict hull of either a Bridgewater Tug or a Preston Brook Tunnel Tug (both of which were similar in design) awaiting conservation at one of the 'wides' near Broken Cross on the Trent & Mersey Canal.

The Bridgewater Barge Coronation at a Preston Brook warehouse (now Pyranha Watersports). The Marina was constructed behind the hedges in the background.

An old photograph showing the original London Bridge before it was replaced with the present concrete bridge. The pub is in the background.

The Barton Swing Aqueduct. The towing path has now been removed.

lift at Anderton was the brainchild of the engineer for the River Weaver who was none other than Edward Leader Williams, the engineer for the Ship Canal; the lift was actually designed and built by Edwin Clark. The design that was eventually settled on was a 'swing aqueduct'.

The swing aqueduct and the proposed swing bridges on the Ship Canal were a similar design to the road bridges that spanned the River Weaver (also designed by Edward Leader Williams). The aqueduct would pivot on an island built in the centre of the Ship Canal and would swing, full of water, to allow ships to pass either side. The navigation trough would be sealed at either end prior to swinging the lock gates to conserve water.

The dimensions of the aqueduct are:

Length	71.6m (235ft)
Width	5.5m (18ft)
Depth of water	1.8m (6ft)
Total weight	1,400 tons (800 tons of which is water)

The weight of the aqueduct is supported by sixty-four steel rollers but, when swung, a greased hydraulic ram takes some of the weight off the rollers. The swinging action is achieved hydraulically, being controlled from a tower on the island that overlooks both the aqueduct and the adjacent Barton Road Bridge. The aqueduct was completed in July 1893 and only then was Brindley's original structure demolished in order to maintain through traffic on the Bridgewater Canal. On the northern bank of the Ship Canal, there are the remains of part of one of the buttresses and approach embankments of the original aqueduct in addition to the site of the Barton Road Aqueduct where the road was spanned by another smaller aqueduct. Even though Brindley's original aqueduct was demolished, there is a similar structure opposite the Old Watch House at Stretford. This is the Hawthorn Lane Aqueduct and, although built on a smaller scale, is reminiscent of its larger brother. With Brindley's Barton Aqueduct out of the way, the finishing touches could be made to the Ship Canal in preparation for its opening on 1 January 1894, having cost £14,347,891 to construct.

Overshadowed by the excitement of the Ship Canal's opening was the closure of the Worsley Mines, the original reason behind the canal's construction. The two entrance portals gave access to no less than 74km (46 miles) of subterranean canals serving coalfaces on several different levels. The miners travelled to the coalfaces down access shafts and the mined coal was loaded into wooden or iron boxes or containers. The containers were then loaded into special 'starvationer' boats, so called because of their exposed ribs. They qualify as being the first container boats in existence. The starvationers made their way out of the mines loaded with containers, negotiating the different levels within the mines by means of subterranean locks or revolutionary inclined planes, also situated underground.

The year 1890 saw the founding of Sprinch's Boat Yard at Runcorn along with the straightening of the canal's route adjacent to the yard. Originally, the canal swung around a sharp bend going towards Runcorn and passed next to the 'Big Pool', a natural lake that Brindley took advantage of when constructing the canal. After passing by the pool, the canal took another sharp bend before proceeding towards Waterloo Bridge (the present terminus of the canal at Runcorn). When the boat yard was built, one of the entrance arms to the pool was filled in and utilised as a dry dock and slipway. Access to the pool was retained by the arm north of the yard. The straightening of the bends took the canal to the front of the yard, as it is today. The dry dock and slipway are still in use, being controlled by the Bridgewater Motor Boat Club.

The canal entered the twentieth century uneventfully. Trade was gradually falling and carried on falling to such a low level that in 1939 Brindley's original line of locks at Runcorn were closed through lack of use; they were filled in 1948. Throughout the early twentieth century, the canal between Worsley and Leigh had been continually affected by mining subsidence. This lead to a need for the canal banks to be reinforced on a grand scale, the work being funded by the organisation responsible for mines at that time, the National Coal Board.

In 1952, new life was injected into the canal by allowing pleasure craft to use the canal for leisure purposes. Boating clubs started to spring up along the canal, the first of which was the Bridgewater Motor Boat Club at Runcorn. Today, their clubhouse stands on the site of Sprinch's Boat Yard, which burnt down and closed in 1935. Other cruising clubs are situated at Lymm, Sale, Stretford (The Watch House), Worsley and Preston Brook. Today, most of the clubs have moorings at more than one location usually of the linear variety, along the side of the canal.

A more recent view of the aqueduct in 1987, when it was painted red and white.

Commercial traffic continued to diminish and, in 1966, the new line of locks at Runcorn was filled in, leaving this end of the canal to terminate at Waterloo Bridge. The canal was severed here in preparation for the building of the new Runcorn Expressway, part of a road network essential to the development of Runcorn New Town. A little thought from the developers would have enabled the line of the canal to be retained for possible future restoration.

The site of the old line of locks is still traceable today. The area had been landscaped and a footpath passes through the centre of what were the once busy lock chambers. However, when the Mersey Second Crossing (the new Runcorn-Widnes Bridge) is completed, the approach roads to the original bridge will be realigned, allowing the canal to be reopened beneath Waterloo Bridge and connect with the Manchester Ship Canal adjacent to Bridgewater House (now a College).

Bridgewater House once started over a forlorn and overgrown wasteland. Today it is a college and is surrounded by a development area. The original lock gates that were removed from the canal when it was infilled had a new lease of life when they were utilised on the restoration of the Upper Avon. Maybe it will not be too long before the canal at the side of it is in use once more.

There is a growing amount of interest and support for the restoration of the new line of locks, allowing access at the Runcorn end of the canal to the Manchester Ship Canal, Runcorn Docks and the Runcorn and Weston Canal, which connects with the River Weaver. Although this restoration would be a formidable task, it would not be an impossible or impractical project,

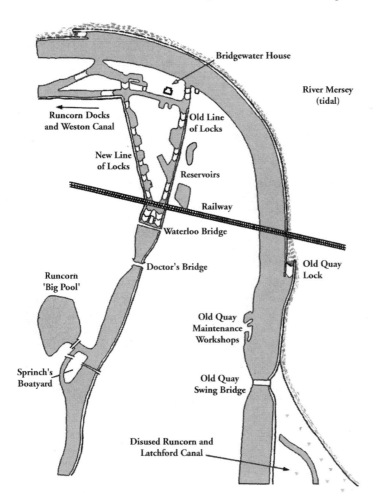

Runcorn area prior to infilling.

A classic photograph of Barton Swing Aqueduct, illustrating how it was swung full of water. Note the horse on the towpath, the barge in the trough and the man in the foreground. Apparently, there weren't any Health and Safety at Work laws then!

An aerial view of Sprinch's Boatyard. The canal swung around the back of the boatyard and entered the Big Pool (top).

The boatyard as it is today.

bearing in mind such projects as the proposed canal connecting Liverpool's Albert Dock to the Leeds & Liverpool Canal, the construction of the Ribble Link on the Lancaster Canal, and the restoration of the Anderton Lift, Standedge Tunnel, and the Huddersfield Narrow and Rochdale Canals.

With the depletion of commercial traffic on the upper reaches of the Ship Canal, the reinstatement of Runcorn Locks could create two new cruising routes for pleasure craft. One, up the Ship Canal to Manchester and the other, downstream to the River Weaver, allowing access to this waterway denied by the long term restoration of the Anderton Boat Lift which was re-opened at Easter 2002.

Apart from creating new bridges and closure of the previously mentioned locks, the Runcorn Expressway also necessitated the filling-in of the Big Pool. Although part of it still remains, it is not connected to the canal except by a small drainage culvert. With a little foresight, it could have been converted into a marina for pleasure craft, but sadly, this was not to be. One of the arms of the canal that gave access to the Big Pool is used by BMBC for off-line moorings.

In 1971, a breach occurred on the embankment over the valley approaching the River Bollin Aqueduct. As a result, part of the embankment and aqueduct were swept away. A new channel was constructed using concrete and steel. The remaining part of the aqueduct was also reinforced and remedial work closed the canal to through traffic for two years.

Commercial traffic ceased on the canal in 1974. This coincided with the opening of a new marina at Preston Brook, opposite the old Norton Warehouses. Preston Brook Marina offers secure moorings off the main line for over 300 craft. There is an exclusive housing development on the canal opposite the marina known as 'Marina Village', which successfully integrated the canal into an urban development scheme and was the first of many such projects to line the canal.

The occasional commercial narrowboat can still be seen on the canal today delivering coal, not to a factory or power station, but for domestic consumption. These floating coalmen are greatly appreciated by boat owners whose craft are fitted with solid fuel stoves and enjoy boating out of season.

Although not strictly part of the Bridgewater Canal, a section of Preston Brook Tunnel collapsed in November 1981. A large crater appeared adjacent to the post office above the tunnel. The formation of the crater sent many tons of debris falling into the tunnel below and damaged a 37m (121ft) long section of the tunnel bore. The post office was damaged beyond repair and demolished, the crater was filled in and the damage to the tunnel was repaired using

A horse-drawn narrowboat at an unidentified location, probably Dunham Town Bridge. Judging by the motor car by the bridge, this photograph was taken in the 1920s or '30s.

concrete sections. The site of the crater is marked on the surface by a round inspection shaft, the size of which can only be appreciated by looking upwards whilst passing through the tunnel. The tunnel was re-opened in April 1984, thus ending two and a half years of isolation for the southern end of the Bridgewater Canal and the northern end of the Trent & Mersey Canal.

In 1984, Bridgewater Estates were purchased by Peel Holdings plc, the property development company. They also purchased the Manchester Ship Canal Company shortly afterwards. The traffic on the upper reaches of the Ship Canal had already diminished, leaving acres of quayside and warehouse land ripe for development. One of their first projects was the construction of an office development at what is now Salford Quays, followed by the Trafford Centre. The latter is a prestige shopping centre built adjacent to Trafford Park. It features 1.4 million square feet of shopping area, 280 shops, thirty-five restaurants, a twenty screen cinema and car parking space for 10,000 cars (but no boats). Peel Holdings plc also own Liverpool's John Lennon Airport and Doncaster's Finningly Airport.

Today, the Bridgewater Canal is much the same as it was over two hundred years ago. There have been obvious changes throughout that period, most of which have been previously documented. The connection with Manchester Docks via Hulme Lock was replaced in 1995 with the new Pomona Lock. The new lock gives access to the Pomona Number Three Dock and is the only lock on the Bridgewater Canal (unless the Runcorn Locks are restored). On the island formed between the Hulme Locks Branch and the main line of the canal, a modern residential building development was completed in 2006. Castlefield has undergone a radical change since Peel Holdings took over responsibility for the canal. Some of the old warehouses have been demolished to make way for modern amenity buildings, such as the YMCA and new theme pubs. In a more dramatic vein, the Middle Warehouse has been completely refurbished and is now used for upmarket apartments, flats and offices. The building also houses Manchester's 'Key 103' and '1152' radio stations on the ground floor. The wharf area has been landscaped and new bridges built to continue the towpath around to areas previously inaccessible. Grocer's Warehouse has been partially rebuilt and houses a replica of the waterwheel and winch driven by the River Medlock.

One of the first boat rallies on the Bridgewater Canal in the early 1950s at an unidentified location.

Unloading barges at Barton Power Station.

An aerial view of Preston Brook Marina taken in the 1990s.

The breach at the River Bollin Embankment and Aqueduct in 1971. Note the three people standing in the centre of the photograph.

The junction of the Old (left) and New (right) Lines of Locks in Runcorn taken from the railway viaduct. The Manchester Ship Canal can be seen in the background.

The same area as it is today looking upwards from the bottom of the Old Line of Locks.

The now disused Hulme Lock.

Pomona Lock, which replaced Hulme Lock.

Developments have taken place at Salford Quays and along the whole of the Manchester Docks complex; the warehouses and factories have been replaced by buildings such as the Lowry Centre and the new Imperial War Museum – North. Looking towards the future, there are plans to build on the area adjacent to Pomona Dock and to convert Pomona Dock itself into a marina. These developments could not be further away from the previous occupants of the area as the theme of the docks has shifted from industry to business, leisure, tourism and housing. There may also be a development of the canal adjacent to the Trafford Centre when the extension to the Metrolink line has been completed. The Bridgewater Canal will host the 2005 National Waterways Festival for the second time in twenty years. The Festival was held at Castlefield in 1988 and Preston Brook was the location of the 2005 rally.

The Bridgewater is a deep, broad and (virtually) lockless waterway that offers a breathing space for the boater before ascending the Rochdale Nine, Wigan Twenty-One or Heartbreak Hill (the name given to the climb from Cheshire into Staffordshire) on the Trent & Mersey Canal. The canal is full of contrasts, from Manchester to Dunham Massey, or from Runcorn to Walton. These contrasts are enjoyed by many thousands of people each year and add to the unique character of Britain's first true canal – the Bridgewater Canal.

A resourceful boater's solution to the problem of where to park the car when boating.

The northern entrance to Preston Brook Tunnel. Compare this shot to the one on page 17 taken of the same location some fifty years previously.

Looking south from the top of the tunnel entrance.

Two

A Description of the Canal's Route

The northern portal of Preston Brook Tunnel marks the beginning of the Bridgewater Canal. The tunnel is 1,133m (1,239yd) in length, and navigation is controlled at all times by a timetable as a result of being unable to see one end from the other. Northbound entry is from twenty past the hour until half past the hour whilst southbound entry is from ten to the hour until the hour.

A little way along the horse path across the top of the tunnel is a pub appropriately named the Tunnel Top (formally the Stanley Arms) which cooks excellent meals in addition to selling alcohol. Opposite the pub is the new ventilation and access shaft, built when the tunnel collapsed in 1981. Soon after emerging from the tunnel, adjacent to the site of the demolished Cotton's Bridge was Preston Brook Station, although no remains are left today. The Old Number One was one of the few remaining warehouses at Preston Brook. In its post-warehouse role, it has been a nightclub and a restaurant but is now rebuilt as a prestige housing project. It lapsed into its previous condition after being destroyed by fire. The warehouse has also featured prominently in an episode of Granada Television's series, *Travelling Man*, which was almost exclusively shot on the Bridgewater Canal.

Claymoore Navigation, adjacent to the A56 road bridge, offers boat hire and the usual boatyard facilities. Opposite are moorings conveniently placed for the Spar supermarket a little way up the road. Further up the road is the Preston Brook residential hotel and conference centre.

Between the A56 road bridge and the M56 viaduct are more reminders of Preston Brook's importance as a canal port in former years. The most prominent reminder is the row of canal cottages used by the people who worked in the numerous warehouses that lined the canal. Overnight mooring is not recommended here due to the close proximity to the railway hidden in a cutting to the left and the M56 motorway above. Nestling in the shadow of the M56 viaduct is Midland Chandlers, whose showroom is a veritable 'canal supermarket' for those interested in apparatus for canal boats.

Immediately after the motorway viaduct is a junction left for Runcorn and straight on for Manchester. Preston Brook's 300 berth marina can be seen on the opposite side of the valley.

Compare these two views of the same
location separated by about eighty-five years.

An echo from the past: the front door of one of the canalside cottages at Preston Brook.

Preston Brook, showing the new Midland Chandlers showroom beneath the M56 viaduct in the background.

The aqueduct over the railway at Preston Brook. Note the partially constructed housing development in the background.

Also on the opposite side of the valley the Runcorn Arm runs parallel to the main line of the canal for a mile or so before turning into an area lined with new housing developments. Looking straight ahead, the white concrete tower of CCLRC's (Council for Central Laboratories Research Councils) Daresbury Laboratories can be seen poking up through the trees. The tower is a well known landmark in this area and, no matter how inviting the canal banks adjacent to the laboratories look, there is no mooring allowed.

Past the laboratories, under Moorefield Bridge, are moorings convenient for visiting the nearby village of Daresbury. The walk up the hill to the village allows a closer look at the tower mentioned earlier. It was originally part of a Van de Graaff generator or, to be more exact, a Vertical Particle Accelerator. This particular project has come to an end and other departments within the laboratories now use the tower. Beneath the ground, in front of the tower, is buried a 'Synchrotron'; this is a large unit used for the research of particle physics and is one of the largest of its type in the country.

The name 'Daresbury' is derived from an ancient word meaning 'oak' and there is certainly an abundance of oak trees in the area. Within the village is a beautiful stone church, of which Reverend Charles W. Dodgson was one of the vicars. Reverend Dodgson took a particular interest in the well-being of the bargemen who crewed the boats that operated on the canal. He was also a friend of Lord Francis Egerton.

One day, when they were both out walking, Reverend Dodgson expressed a wish to provide the bargemen with some sort of religious privileges. 'If only I had £100', he said, 'I would turn one of those barges into a floating chapel.' He went on to describe to Lord Egerton how he would have the boat fitted out and how he would cruise along the canal giving religious instruction and holding services. A few weeks later, he received a letter from Lord Egerton, informing him that his floating chapel was ready and waiting for the reverend to collect it from Worsley. This floating chapel is believed to be the first of its kind and was used regularly by people who worked along the canal.

Due to the width of the Bridgewater Canal, a crane has to be used when lifting the stop-planks into position.

George Gleave's Bridge near Daresbury, hidden in the foliage.

Thomason's Bridge near Moore.

The floating chapel wasn't Reverend Dodgson's only claim to fame. He was father to the renowned Victorian photographer and author, Charles Ludwidge Dodgson, better known by his pseudonym of Lewis Carroll, and famous for writing *Alice's Adventures in Wonderland* and *Alice Through the Looking Glass*. If the church in Daresbury is visited, the stained glass windows are of interest as they depict characters from the *Alice* books. There is also an appreciation centre in the village, conveniently placed next to the Ring O'Bells pub where thirsts can be quenched and stomachs filled before the return walk back to the canal.

After the two Daresbury bridges, at the next bridge, Moore Bridge, is a track that leads to the village of Moore. There is a general store and the Red Lion pub, the illuminated sign can be seen from the canal beckoning the thirsty. They also serve meals of a high standard.

Around the corner, just before Acton Grange Bridge, is a memorial erected by the locals to the memory of 'Ken the Tramp'. Ken was a friendly, intelligent, professional man who dropped out of society of his own will. He could usually be seen huddled beneath his umbrella and was always ready to catch the ropes and have a chat over a mug of tea until he died of pneumonia in March 1984. He will be remembered and missed by many of the more experienced boaters on the canal.

The Red Lion can also be reached from Acton Grange Bridge. There are craft moored just after Acton Grange Bridge on private moorings. This stretch is also a handy mooring for the convenience store and post office, 150m along the road.

Between Moore and Walton, the canal winds through delightful countryside punctuated by a wooded cutting. Here there is a good sheltered mooring, perfect for barbecues.

Before long, Walton's A56 Chester Road Bridge can be seen. This bridge is the last surviving concrete 'Toast Rack' bridge on the Bridgewater Canal. Just before and after the bridge are moorings of the Bridgewater Motor Boat Club, based at Runcorn, opposite which, are excellent overnight moorings. At the next bridge, the excellent Walton Arms pub can be reached by going down the road on the towpath side. From the A56, near the Crematorium entrance, can be seen one of the two remaining transporter bridges in the country. This example is scheduled as a listed structure and spans the old Mersey & Irwell Navigation, adjacent to the Crossfields

The memorial garden and stone for Ken the Tramp at Moore.

When Chester Road Underbridge sprang a leak in 1987, flexible plastic sheeting isolated the affected area in order for remedial work to be undertaken.

Chemical Complex (the original Lever Brothers factory) in Warrington. Today it is owned by Warrington Town Council.

At the next bridge, Walton Lea Bridge, care must be exercised when passing other craft due to the jagged rocks jutting out from the canal banks. Once at the bridge, steps give access to Walton Park. The park is well worth a visit and possesses children's swings, a children's' zoo, beautiful gardens and crazy golf. The house itself is a function and conference centre in addition to having an appreciation centre, café and sweet shop.

After the cool darkness of the cutting, the canal winds its way to Stockton Heath, with the trim back gardens and tidy houses on one side of the canal overlooking the rolling Cheshire countryside on the other. Stockton Heath was once the site of Stockton Quay, a busy trans-shipment point in the canal's heyday. The old quay contained many warehouses in addition to Cash's Boatyard, the basin for which can still be seen in the shape of a square winding hole. The yard and warehouses were demolished in the mid-1980s to make way for the 'Old Quay' housing development.

The London Bridge pub was also one of the points along the canal where the passenger flyboats, such as the previously mentioned *Duchess Countess*, would stop for a change of horses and also to allow passengers to embark and disembark. The actual point where the boats stopped was at the steps next to the bridge, adjacent to the pub. The London Bridge is a favourite with canal boaters, offering meals as well as drinks. Boats can moor next to the old flyboat steps whilst visiting the pub.

Opposite the Old Quay development is Thorn Marine. They offer all the usual boat shop facilities such as water and sanitary disposal and is a veritable 'Aladdin's Cave' for chandlery, boat parts and canalia. The left hand side of Thorn Marine's shop used to be the Lancashire Constabulary (and the canal's before it) mortuary, in use until 1951, the tiles of which can still be seen to this day. Rubbish disposal is at a recycling centre, access to which can be gained via a gate 100m along the towpath from Thorn Marine (opposite the small marina moorings). As well as a garage and shopping centre there is an off-licence, which is reached down the road at the side of Thorn Marine. A little further on down the side road, on the main A56 road, there

A view of Walton Road Bridge that boaters do not usually see. Note the toast-rack design, the only surviving example on the Bridgewater Canal.

is a convenience store that is open 'eight till late' and an excellent Chinese takeaway. Across the Manchester Ship Canal swing bridge is a large Morrison's supermarket.

Most of the original canal buildings have disappeared although some still remain, such as the Bank Rider's House (Thorn Marine), the Smithy and Stables (behind the London Bridge pub) and a couple of warehouses (behind the new housing development). Another reminder of the canal's past is on a plaque at the top of the flyboat steps adjacent to the pub. Around the back of Thorn Marine, a mural has been painted on a galvanised steel fence, which adds a bit of colour to the area.

Next to the pub is a small marina that only offers moorings to residents of the adjacent buildings. After passing the back of more houses, the canal strides across a small valley on an embankment adjacent to another new housing development. A side road passes beneath the canal on a small aqueduct or 'underbridge' as they are called on the Bridgewater.

The next port of call is Grappenhall where the A56 runs alongside the canal for quite some distance. There are private moorings at Stanney Lunt Bridge and, through gaps in the towpath hedge, are bus stops on the A56 that offer a regular bus service to Warrington. Nearby are shops and a post office that can be reached from Cliff Lane Bridge. Care must be exercised at the bend just after Grappenhall Bridge. It is an acute, blind bend and Sod's Law dictates that another craft will be met here.

The canal leaves civilisation behind for a while until Thelwall is reached. There are many boats moored on two private linear moorings just past Knutsford Road Bridge (A50) and Pickering's Bridge. Just before Thelwall Underbridge is reached, logs for the solid fuel stove can sometimes be purchased from the shed on the offside of the canal. A short way along the canal there are good moorings in the shelter of trees, ideal for a barbecue or children to explore. After the canal emerges from the wooded cutting, Thelwall Viaduct lifts the M6 motorway high above the Bridgewater Canal, the Manchester Ship Canal, the old Mersey & Irwell Navigation, a railway and a road on lofty concrete pillars. It would be interesting to know James Brindley's thoughts on this particular piece of civil engineering had he been able to witness it. There is

Two shots of the canal at Walton. Note how the design of pleasure craft has changed in the twenty years separating the photographs.

Runcorn's Bridgewater Motor Boat Club moorings at Walton.

The transporter bridge over the Mersey & Irwell Navigation near Walton.

Two views of the wharf at Stockton Heath, separated by fifteen years. The first view is from the location of Cash's Boatyard.

a garage that sells milk, bread and other staples at Ditchfield Bridge before which are more moored craft on private linear moorings.

A new, tastefully designed housing development heralds the approach to Lymm. The town of Lymm itself is soon reached after negotiating a small rock cutting. This is a beautiful town, full of character. As well as many pubs and shops, there are medieval stocks and a cross from the same era worthy of inspection plus an unusual linear park. The park is set in a gorge that runs right through the town centre. The stream that runs through the gorge emanates from a large lake known as Lymm Dam that is popular with anglers, walkers and picnickers. The stream and the adjacent road pass beneath the canal adjacent to Lymm Green Bridge. There are temporary moorings on both sides of the canal, and on the offside of the canal is a car park that accommodates a market every Thursday. There is an excellent fish and chip shop and a good Chinese takeaway opposite the post office in the side road on the offside of the canal as well as many pubs.

Just before the car park is the entrance to a small, disused tunnel that once housed the icebreakers used to keep the canal open for traffic during the winter. These craft had a rounded bilge and were specially sheathed in iron to prevent the ice cutting into the timber of the hulls over a prolonged period of time. They were pulled by a team of horses whilst the crew members hung onto a central bar and rocked the boat from side to side. As the boat progressed along the canal, it opened up a channel, allowing other craft to travel along a previously ice-bound waterway. One humorous tale is of an ice breaker that rose onto the ice. It was pulled at a great speed, skidding along the ice, by the horses for quite a distance before they could be brought under control and the business of icebreaking resumed.

Immediately after Lymm Green Bridge there used to be a fine example of a typical Bridgewater Canal warehouse that featured the original lifting equipment used for hoisting goods in and out of the boats. Unfortunately, the warehouse has now been demolished and a new housing development now occupies the site. Just around the corner is the headquarters of Lymm Cruising Club and more of their extensive moorings. Adjacent to their slipway, there is a small arm that used to lead to an unloading tunnel now euphemistically called Lymm Tunnel.

The Icebreaker Tunnel at Lymm.

Icebreaking on the Bridgewater Canal. The ropes leading from the boat were attached to the horse.

As the canal leaves Lymm behind, excellent views of the rolling Cheshire countryside can be seen. More boat moorings from Lymm Cruising Club are seen around the bend from the Lymm straight at Oughtrington. The bridge used to be known as Oughtrington Bridge but has been renamed Lloyd Bridge after a family that served the Bridgewater Canal for many generations. Immediately after the bridge, at Oughtrington Wharf, Calor gas can be obtained.

The canal now winds into the Agden area. Here, there is another excellent example of an underbridge. Immediately after Agden Bridge is a warehouse that was once used as a hospital for canal horses, part of which has been tastefully converted into a private house. The original winch and entrance for the horses can be seen on the canal side of the building. This building is at the beginning of a long straight stretch of canal on which is situated three boatyards next to each other. Northern Marine Services is first, next to which is the Barn Owl pub and restaurant. There is no provision for mooring at the Barn Owl so customers are requested to moor on the towpath side of the canal and ring a bell for a ferry boat to collect them and return them to the towpath after being fed and watered. This is followed by Hesford Marine and Lymm Marina (formally Ladyline). The two latter boatyards have excellent chandleries and every service that may be required ranging from Calor gas, pump-outs, diesel fuel, moorings, boat sales and repairs whereas Northern Marine Services are boatbuilders only.

More linear moorings for Lymm Cruising Club are passed and soon the Old Number Three pub is reached. There is a water tap and temporary moorings adjacent to the footpath that leads to the pub. This is a good point to meet up with visitors coming by car as the A56 runs at the front of the pub where there is a convenient lay-by. As well as the temporary moorings there are also permanent moorings.

Around the bend from the Old Number Three, the canal is built on an embankment before it crosses the River Bollin on an aqueduct. This was the site of a major breach in 1971 that closed the canal to through traffic for two years whilst the aqueduct and adjacent embankment were rebuilt. On the offside of the canal can be seen a mill that harnessed the River Bollin to drive a large waterwheel which, in turn, drove the mill machinery. The waterwheel has gone but the head and tailraces are still in existence and are well worth a visit. Just after the embankment are more temporary moorings.

Dunham Massey is soon reached with its old school, deer park, stately home and gardens. It is well worth mooring to spend an afternoon at the grand house and deer park. The Axe & Cleaver pub is a favourite with boaters as is the Bay Malton a mile further on. Close to the Axe & Cleaver is a post office and general store. After winding around in true contour fashion for a mile or so the Bay Malton pub is reached. For those members of the crew with an excess of energy due to the lack of locks, a disco is held here on certain nights of the week.

A view of the canal at Lymm in 1986. Apart from the design of the cabin cruiser, the scene would be identical today.

The Old Canal Horse Hospital near Agden on the outskirts of Lymm.

Lymm Cruising Club's moorings at Agden.

The flooded River Bollin as it passes beneath the rebuilt aqueduct at Dunham Massey.

The Old Watch House at Stretford.

The canal now starts to change its character. It leaves behind the rural atmosphere and becomes increasingly industrial as factories line the canal. A new marina development and the old Linotype works heralds the approach to Altrincham. The old Linotype works used the Bridgewater Canal for both receiving raw materials and distributing finished products. This is confirmed by the extensive wharfing facilities that were once situated here. As well as Linotype, there was a major coalyard belonging to Bridgewater Estates where the new office block now stands. Altrincham has a busy town centre and offers all the amenities and shops that one would expect.

Before long, the railway joins the canal and is its constant companion until Stretford. Although originally a railway line, Manchester's Metrolink trams run along the line and offer a regular service straight to the heart of Manchester and the surrounding area from the stations situated at regular intervals along the canal. A seemingly unending straight stretch, with similarly unending linear boat moorings belonging to members of Sale Cruising Club on the offside of the canal, signals the approach of Sale. The town centre is reached through a hole in the wall just after Sale Bridge. The town centre offers a large precinct with every conceivable shop plus a couple of garages for petrol, oil and other services.

At the end of the straight stretch is a pub that caters for boaters called, ironically, the Railway and it is adjacent to Dane Road Bridge. The canal leaves Sale behind and winds around for a bit before passing the boathouse of Manchester University Rowing Club. Soon after this, the infant River Mersey is crossed on the Barfoot Aqueduct. The M60 motorway also crosses the canal here and marks the beginning of another straight stretch lined with craft, this time belonging to members of the Watch House Cruising Club. The headquarters for this society is a very distinctive piece of canal architecture. As the name of the club implies, it was a watch or lookout post. In the days of passenger transport on the canal, when a fly-boat was spotted in the distance, a fresh set of horses was made ready to delay the boat as little as possible when it stopped to change horse teams. Cutting through the side road from the Watch House gives access to the Stretford Arndale Centre. Here there are many shops that will meet the needs of even the most dedicated shopper.

The Watch House signals the end of Sale Moor and the start of Urmston on the outskirts of Manchester. Urmston has a shopping centre reached from Edge Hill Bridge. The canal is now very industrial and care must be taken to avoid tyres, plastic sheeting, mattresses and other waterborne debris.

After winding through Urmston, the canal reaches a T-junction called 'Waters Meeting'. To the left is the Leigh Branch, which was the original part of the Bridgewater Canal leading to Barton Aqueduct, Worsley and the end-on junction with the Leeds & Liverpool Canal at Leigh. This branch is described later but for now we will turn right and head into Manchester and the junction with the Rochdale Canal at Castlefield, the canal's terminus. This stretch is also part of Brindley's original canal and originally passed through a canyon lined with factories and warehouses.

Originally one of the busiest ports in the country, the docks are now mainly used for leisure although the occasional Mersey Ferry and coastal traffic can be seen in the complex. Manchester United's football ground dominates the skyline for the last mile or so and the canal runs next to the rear of the ground. Soon after this, the canal's towpath changes sides at Throstle Nest Bridge immediately before a re-profiled stretch of canal is reached. Pomona Station is right on the canal banks and would be a suitable place to moor the boat and catch the Metrolink into Manchester if the remainder of the canal is not to be navigated.

The River Irwell can be glimpsed through the supports of the Metrolink viaduct. Access to the River Irwell and Manchester Docks is achieved by the new Pomona Lock, constructed to replace the original Hulme Locks, the site of which can be seen a little further on past Woden Street Footbridge where the River Medlock joins the River Irwell. If access to the docks, River Irwell and Manchester Ship Canal is required, please refer to the section dealing with navigational information. Close to the Pomona Lock, the canal is criss-crossed by the railway and Metrolink. A little further on from Pomona Lock, on the towpath side of the canal, can be seen an early example of a circular Brindley overflow weir. This example may have been a prototype for the type seen on the Staffordshire & Worcestershire Canal.

The River Irwell is fairly wide at this point and, originally, boats could only navigate downstream as far as Woden Street Footbridge, where the River Irwell flowed into Manchester' Docks. Boats can now navigate through the docks to the start of the Manchester Ship Canal, which is marked by Lowry Footbridge, but stringent preparations must be made prior to navigation. Passage along the Ship Canal is controlled due to commercial seagoing vessels using the lower reaches of the canal. Upstream, the river passes through the centre of Manchester to the head of navigation at Hunt's Bank.

Cut Hole (Hawthorn Lane) Aqueduct at Stretford. This is a scaled-down version of the original Barton Aqueduct

Compare these two shots of the canal at Old Trafford near Manchester. They were taken from almost identical viewpoints and illustrate how much the canal has changed in recent years.

The Metrolink spanning the canal at Pomona Lock.

Pomona Lock.

The upper Hulme Locks before the re-profiling of the canal.

The entrance to the disused Hulme Locks.

The entrance to Potato Wharf at Castlefield.

Besides the Bridgewater and Manchester Ship Canals, there were two other canals that joined the River Irwell. They were the Manchester, Bolton & Bury Canal, the lower parts of which are filled in but traceable (just), and the Manchester & Salford Junction Canal. The latter was basically a tunnel connecting the Rochdale Canal to the River Irwell in an attempt to prevent the Bridgewater Canal's monopoly on passage to the River Irwell via Hulme Lock. Only isolated glimpses of this canal can now be seen although the remains of a lock chamber are still present beneath the Granada Television Studios, as are some of the lengths of tunnel, which were once used as air raid shelters during the Second World War.

Returning to the Bridgewater Canal, soon after passing beneath Egerton Street Bridge, the cast iron railway viaducts that surround Castlefield Junction can be seen. To the left is Potato Wharf where the remains of Brindley's original cloverleaf overflow weir can still be seen. Straight on is the junction with the Rochdale Canal, which can be reached via Duke's Lock. To the right are the extensive wharves and the branch that leads up to the River Medlock and the canal's terminus.

The Castlefield area was tidied up and dredged prior to the 1988 Inland Waterways National Rally of Boats. Since then a new footbridge from the beginning of the Rochdale Canal, which spans the entrance to the Castlefield Basin, has been constructed. There was also considerable development of the site in recognition of its industrial heritage. This development included the construction of additional bridges, converting warehouses into residential apartment buildings, theme pubs, and even a radio station. The buildings not renovated have been demolished to make way for developments more in keeping with the canal's new image.

The Rochdale Canal was previously owned by the Rochdale Canal Company but now comes under the jurisdiction of British Waterways. A special licence was required to cruise the Rochdale Canal, but since coming under the umbrella of British Waterways, the British Waterways licence is required.

Manchester City Centre is reached by going across the new footbridge and around the wharf to Deansgate where the shops and major attractions are a fifteen-minute walk to the left. A little way along the road is the entrance to the Manchester Museum of Science & Technology. For a small entrance fee, displays ranging from steam engines, aircraft and road transport can be seen in addition to modern technology exhibits. Also worth a visit are the remains of the Roman Fort that gave Castlefield its name. There was also a Roman fosse which was an early form of navigable cut that connected the Rivers Irk and Irwell. Unfortunately, nothing remains of this early navigation. Over the River Irwell from the Granada Television Studios is the Mark Addy pub. The pub is named after a man who, as a seven year old boy, was instrumental in rescuing an oarsman from the river close to Albert Bridge. In later life, Mark Addy saved more than fifty people from drowning.

In addition to the obvious attractions mentioned earlier, there are many interesting canal features in the area. At the end of the arm adjacent to Deansgate is the River Medlock. It plunges into a syphon, or tunnel, and runs beneath the length of the arm to emerge behind Potato Wharf and Brindley's cloverleaf weir. At the other side of the bridge, at the end of the arm, can be seen the rebuilt entrance to the unloading tunnel. A chamber beneath the canal contained a waterwheel turned by the Medlock, which was the motive power for the winch that raised the coal and other cargoes in containers from the canal boats up to street level. The remains of the other end of the tunnel can be seen at Pioneer Wharf, adjacent to Deansgate, on the Rochdale Canal. When this canal was built it cut across the line of the tunnel, which effectively shortened it. Further details, the latest attractions and a road map can be obtained from the Heritage Centre adjacent to Deansgate.

Looking from Potato Wharf at Castlefield towards the Merchant's (on the left) and the Middle Warehouse (centre).

Looking down the Castlefield Arm from Grocers Warehouse. These perspectives illustrate the landscaping and general tidying-up that has taken place in this location. About twenty years separate the photographs.

The new footbridge at the end of the Castlefield Arm adjacent to Grocers Warehouse.

Looking up the Rochdale Canal towards Deansgate and Pioneer Wharf.

Above: The Middle Warehouse c.1960.

Below: The same scene today.

Merchants Warehouse, adjacent to the junction with the Rochdale Canal was virtually derelict but is now a prestige office development.

Barges being unloaded adjacent to the Middle Warehouse.

The new Merchant's footbridge carrying the towpath over Castlefield Junction.

Reflections at Castlefield: Potato Wharf and the complex of bridges that span the adjacent refurbished wharves.

A diesel Bridgewater Tug (similar to BW's Bantam Tugs) being launched at Castlefield.

Barton Swing Aqueduct.

74

The Leigh Branch

From Stretford 'Waters Meeting' the canal winds around factories from Trafford Park Industrial Estate. Soon a long straight stretch is reached. The water along this stretch is unbelievably clear. The plastic bags and other submarine obstacles can be easily spotted and avoided. The fish dart about, some of them very big, lurking beneath the weeds. No wonder this stretch is so popular with fishermen.

At the end of the stretch is Barton Swing Aqueduct. Not very long ago, the Ship Canal, spanned by the aqueduct, was busier than the Bridgewater Canal. Now, the tables have turned, and the Bridgewater is the busier route. To the left is the original line of the canal before Brindley's Aqueduct was demolished. If the aqueduct is open to the Bridgewater canal, cruise straight over, but care must be exercised if it is windy. During the winter months, the aqueduct is periodically closed for maintenance. The Manchester Ship Canal Company should be consulted for confirmation of the closure dates.

Once over the aqueduct, the canal passes through Patricroft where there are shops and a garage. The moored boats belong to members of Worsley Cruising Club whose clubhouse is adjacent to an old loading wharf. The canal soon swings beneath the Liverpool & Manchester Railway line, which crosses the canal on a bridge reminiscent of the original Barton Aqueduct in design. From Barton the canal is very weedy. When cruising the canal, it is advisable to have a towel ready to dry your hands just in case de-weeding is necessary.

The water now starts to have an orange tint to it, which grows stronger as Worsley is approached. Worsley has much to offer. There are shops, a pub, sanitary station, water point, dry docks, hire boats and the mines. A new water treatment plant completed in 2005 was constructed to purify the orange water draining from the mines in an attempt to minimise pollution. Down the road is a garage and Calor gas dealer, plus a regular bus service to Manchester. Opposite the sanitary station is the boathouse that once sheltered the *VIP* Barge. Adjacent to the moorings is the sanitary station and water point. A walk across the road will be rewarded by the opportunity to have a drink in the Bridgewater Hotel pub.

The Packet House at Worsley Turn. The entrance to the Delph is under the bridge to the right.

Adjacent to the Old Boat House, the old oil stores have been tastefully transformed into private apartments. There are other places of interest such as the Duke of Bridgewater's original mine entrances. The short branch to the mines is no longer navigable and starts at the Alphabet Bridge, next to the Packet House steps. This bridge is so called because the walkway has twenty-six planks across its span. Children used to sing and practice their alphabet when crossing the bridge by reciting a letter per plank. To reach the mine entrances, walk down the path signposted 'The Delph' at the side of the Casserole Restaurant. The Delph is a large basin with two arms leading to the mine entrances. There is also a sunken 'Starvationer', the first known type of container boat, kept in the basin. Other examples of this type of boat and the associated containers can be seen at the Ellesmere Port Boat Museum.

Boothstown Pen when it was used as a graveyard for boats.

The same location today houses a marina, canalside housing development and The Moorings pub.

Leaving Worsley beneath Worsley Bridge, the canal is crossed by the M62 and the M60 Manchester outer ring motorway, after which Worsley Old Hall, the Duke of Bridgewater's residence, can be seen through the trees. The Hall is now used as a conference centre. The orange colour of the canal, caused by the drainage from the mines, soon starts to fade. It is always possible to tell when a boat has visited Worsley by the orange stain left by the water, but it washes off easily. On the left can be seen the remains of the old Hollins Ferry Branch, now filled in and last used as a dredgings dump.

The next point of interest is the Boothstown Pen. Here, there were more underground canals, but only one and a half miles compared to Worsley's forty-six. It was also the terminus of the Bridgewater Tramway, which transported coal from outlying mines to the canal. The basin was once surrounded by warehouses, now demolished, and was quite a busy area in its heyday. In later years, after the mines had been closed, the basin became a boats' graveyard where fishermen tried to lure fish from their hiding places between the sunken narrowboats, Leeds & Liverpool shortboats and Bridgewater barges. The entrance from the canal was filled in to prevent boats being damaged on the many sunken craft. In 1989, the basin was emptied, the old boats removed and the basin converted to off-line marina moorings complete with a small canal shop. An upmarket housing development was built, the surrounding area landscaped and a new pub, The Moorings, added. Whilst looking at this area, it is surprising just how clear the water is in direct contrast to a short distance away at Worsley.

The canal threads its way through the strange landscape filled with coal tips until Astley is reached. Here there is a mining museum and two pubs, access to which is gained via Astley Bridge. There are also some shops and a post office. Due to the continuing threat from subsidence two new stop gates have been installed to minimise the risk of leakage from the canal. The chimneys and mills of Leigh slowly grow closer. After passing Butt's Basin, mills and a large school, the canal is once more in a factory-lined canyon. There are boats moored at Butt's Basin with a sanitary station nearby at Butt's Bridge. A little further on is located Lorenz Boat Services. The basin is also home to the 'Water Womble', an ex-Leeds & Liverpool shortboat that travels along the canal acting as a waterborne road sweeper. Butts Bridge, Mather and Leigh bridges give access to Leigh town centre which has all the usual amenities.

The best moorings are through the stopboards that mark the end of the Bridgewater Canal and on the Wigan side of Lea Bridge, in the basin opposite the old Leeds & Liverpool Canal warehouses which are now tastefully renovated.

If you are travelling on towards Wigan, a British Waterways Board Officer at Plank Lane Lift Bridge sometimes takes boat details and checks the boat's licences.

The rebuilt Mather Bridge in Leigh.

The end of the Bridgewater Canal at Leigh is marked by the stopboards, where it makes an end-on junction with the Leeds & Liverpool Canal.

One of the old unloading wharves at Preston Brook prior to the building of the housing development.

The Runcorn Arm

Immediately after turning onto the Runcorn Arm from the main line the canal crosses a railway on a modern concrete aqueduct. Just past the aqueduct are the old trans-shipment wharves (which are now private moorings) and Pyranha Watersports Centre which occupies an old warehouse that has been tastefully extended. Pyranha produce fibreglass canoes that are renowned worldwide.

Opposite the Marine Village housing development is the entrance to Preston Brook Marina. The Marina offers secure moorings for over 300 boats as well as brokerage, slipways, Calor gas, boat and engine repairs in addition to winter storage. The Marina is owned by the Manchester Ship Canal Company. 100m past the marina entrance is a water tap, toilet and sanitary station.

At Borrow's Bridge, Runcorn East Station, shops, a pub and a fish and chip shop are a short walk up the hill. Along this stretch of canal, there are excellent views across the valley towards Daresbury. Norton Bridge is a rare commodity on the Bridgewater canal; it is a changeover bridge where the towpath changes sides. The canal now enters a very pleasant wooded cutting as Norton Priory is reached. The Priory is well worth a visit and features beautiful gardens and restored buildings. Access is gained via Green's Bridge and following the path over the bridge. Also at Norton are swimming baths, squash and tennis courts in addition to running tracks. Immediately after Green's Bridge, the canal negotiates an 'S' bend as it skirts a large natural lake. Just before Astmoor Spine Road Bridge, The Barge pub and restaurant is situated.

The towpath reverts to its usual side at Old Astmoor Bridge, which is followed by a wide stretch of canal that was once lined with warehouses. Today, the only clue to their past existence is the occasional stump from a loading crane and the mooring rings. The next stretch is very exposed giving a good view of Fiddler's Ferry power station. This stretch is renowned for the winds coming in from the River Mersey.

Soon boats moored at the Boat and Butty Company are passed. The Runcorn-Widnes Suspension Bridge can be seen on the horizon before warehouses enclose the canal. The Grapes Hotel is adjacent to the footbridge with the Egerton Arms just around the corner. A little

The same location today.

Approaching Greens Bridge at Norton Priory.

further on is the headquarters of the Bridgewater Motor Boat Club, the oldest boat club on the canal, founded in 1952. At one time, the area was Sprinch's Boat Yard, but now BMBC members operate the slipway and dry dock built on one of the old arms that used to lead to the Big Pool. The Clubhouse has a bar and the atmosphere is cordial. Boaters are invited to call for a chat and enjoy their hospitality. The Arm, just past the Clubhouse, (now used for members' moorings) once lead to the Big Pool before the building of the Runcorn Expressway necessitated its infilling.

The end of the canal is now in sight. At Waterloo Bridge there are more moorings belonging to BMBC members with the Waterloo Bridge pub opposite and easy access to the town centre, which is a short distance away. The three arches of bridge once spanned a canal dual carriageway with a dry dock in the centre, but were removed to make way for the Runcorn Expressway. If time allows, a walk down the old line of the canal is a must. The two lines of locks are still traceable down to Bridgewater House, where the canal once entered the Manchester Ship Canal as well as giving access to the Runcorn and Weston Canal, which connected with the River Weaver via Runcorn Docks.

It is a pity the canal had to be vandalised by town planners. When walking around here one thinks how easy it could be to re-open. Access to the Runcorn and Weston Canal would give a second route into the Weaver and create a small circular cruising route in the shape of the 'Runcorn Ring', consisting of the Runcorn and Weston Canal leading to the River Weaver, up the Anderton Boat Lift and onto the Trent & Mersey Canal before returning to the Bridgewater Canal.

With the current upsurge in recognising the potential of our canals in urban development schemes and the re-opening of many disused canals, we live in hope that the mistakes of the past will be rectified in Runcorn as they have at other locations. The development of the canal in this area would open up the isolated northern end of this branch and give it a renewed focus. The Castlefield area was rejuvenated after the 1988 IWA rally, maybe the same will happen to Runcorn after the National Waterways Festival in 2005.

The canal winds around the old boundaries of Sir Richard Brooke's estate at Norton Priory.

The approach to Runcorn, windswept by the breezes coming off the River Mersey estuary.

Once in Runcorn, the canal winds around to where Sprinch's Boatyard used to be.

All that remains of Runcorn's Big Pool. Rumour has it that if you dig down below the water, the remains of narrow boats and barges will be found.

The arm that joined the Big Pool now offers off-line moorings.

The clubhouse of the Bridgewater Motor Boat Club, the first boat club on the canal and founded in 1952.

Waterloo Bridge, the canal's terminus at Runcorn. The canal is infilled beyond this point.

One of the old entrance locks to the Manchester Ship Canal adjacent to Bridgewater House.

Icicles hanging from one of the ventilation shafts for Preston Brook Tunnel.

The old horse path over the top of the tunnel. The horses were led along the path whilst the boats were either legged or towed through.

85

A wooden bridge guard that protected the bridge's brickwork from being damaged by towropes.

The canal at Preston Brook looking towards the tunnel from the A56 road bridge.

Three

Linear Maps of the Canal's Route

The accompanying maps are not to scale but each page shows approximately two miles of canal. The size and scale of some features may have been distorted in the interest of clarity. Not every canalside feature is included, only those that are likely to be of use or interest. In some areas of the canal, especially those undergoing development, it is impossible to include features that have been added since the maps were produced.

Preston Brook Tunnel

Preston Brook Tunnel is 1,133m (1,239yd) in length and is not straight. One cannot quite see from one end to the other. As a result of this, a timetable is in operation for craft movement at all times.

Craft Entry Southbound   Entry is only between twenty and thirty minutes past the hour.
Craft Entry Northbound   Entry is only between ten minutes to the hour and on the hour.

About halfway through the tunnel, the new sections constructed when the tunnel collapsed in 1982 can be seen. A quick look upwards when passing the ventilation duct (drips allowing) will give some idea of how far beneath the ground the tunnel is situated.

Branch Distances

Preston Brook to Runcorn Waterloo Bridge	8.0km	(5.00 miles)
Preston Brook to Preston Brook Tunnel	1.2km	(0.75 miles)
Preston Brook to Stretford	33.0km	(20.50 miles)
Stretford to Leigh	17.3km	(10.75 miles)
Stretford to Castlefield	4.4km	(2.75 miles)
Woden Street Footbridge to Hunt's Bank (River Irwell)	3.6km	(2.25 miles)

The derelict 'Old Number One' overlooks a frozen Bridgewater Canal at Preston Brook.

Mileage Chart

Main Line

Preston Brook Waters Meeting	0km	(0 mile)
Red Brow Underbridge	1.6km	(1 mile)
Keckwick Bridge	3.2km	(2 miles)
Moore Bridge	4.8km	(3 miles)
Walton Bridge	6.4km	(4 miles)
London Bridge	8.0km	(5 miles)
Lumbrook Underbridge	9.6km	(6 miles)
Stanney Lunt Bridge	11.2km	(7 miles)
Pickering's Bridge	12.9km	(8 miles)
Thelwall Viaduct	14.5km	(9 miles)
Barebank Bridge	1.6km	(10 miles)
Oughtrington Bridge	17.7km	(11 miles)
Agden Bridge	19.3km	(12 miles)
Bollin Aqueduct	20.9km	(13 miles)
Season's Moss Bridge	22.5km	(14 miles)
Broadheath Bridge	24.1km	(15 miles)
Timperley Bridge	25.7km	(16 miles)
Marsland Road Bridge	27.3km	(17 miles)
Doctor White's Bridge	28.9km	(18 miles)
The Old Watch House	30.5km	(19 miles)
Longford Road Bridge	32.1km	(20 miles)
Stretford Waters' Meeting	33.0km	(20.5 miles)

The same location today, converted into prestige apartments.

Leigh Branch

Stretford Waters Meeting	0km	(0 mile)
Parkway Road Bridge	1.6km	(1 mile)
Barton Aqueduct	3.2km	(2 miles)
Patricroft Railway Bridge	4.8km	(3 miles)
Worsley	6.4km	(4 miles)
Keeper's Turn	8.0km	(5 miles)
Boothstown Pen	9.7km	(6 miles)
Vicar's Hall Bridge	11.2km	(7 miles)
Morley's Bridge	12.9km	(8 miles)
Marsland Green Bridge	14.5km	(9 miles)
Butt's Basin	16.0km	(10 miles)
Leigh Stopboards	17.2km	(10.75 miles)

Manchester Branch

Stretford Waters Meeting	0km	(0 mile)
Pomona Lock	1.0km	(0.6 mile)
Hulme Lock	3.2km	(2 miles)
Castlefield Junction	4.4km	(2.75 miles)

The frozen Bridgewater Canal at Preston Brook Waters Meeting.

Marina Village overlooks the canal at Preston Brook.

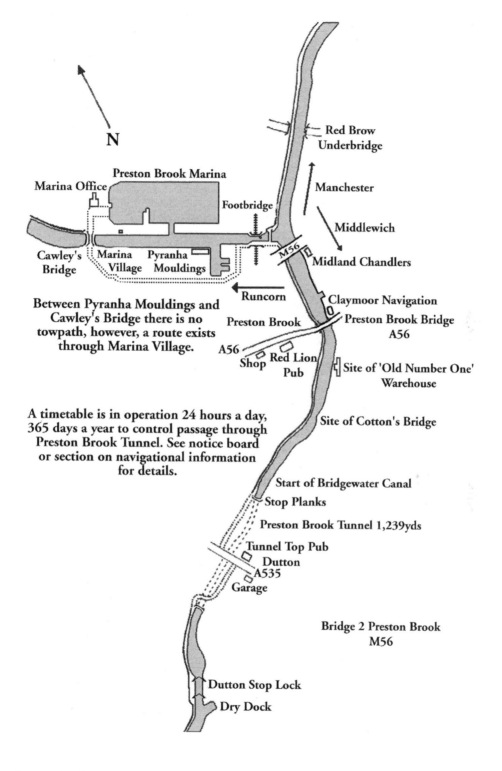

N

Red Brow Underbridge

Preston Brook Marina

Marina Office

Footbridge

Manchester

Middlewich

M56

Midland Chandlers

Cawley's Bridge

Marina Village

Pyranha Mouldings

Runcorn

Claymoor Navigation

Between Pyranha Mouldings and Cawley's Bridge there is no towpath, however, a route exists through Marina Village.

Preston Brook

Preston Brook Bridge A56

A56

Shop

Red Lion Pub

Site of 'Old Number One' Warehouse

A timetable is in operation 24 hours a day, 365 days a year to control passage through Preston Brook Tunnel. See notice board or section on navigational information for details.

Site of Cotton's Bridge

Start of Bridgewater Canal

Stop Planks

Preston Brook Tunnel 1,239yds

Tunnel Top Pub

Dutton

A535

Garage

Bridge 2 Preston Brook M56

Dutton Stop Lock

Dry Dock

Preston Brook Area

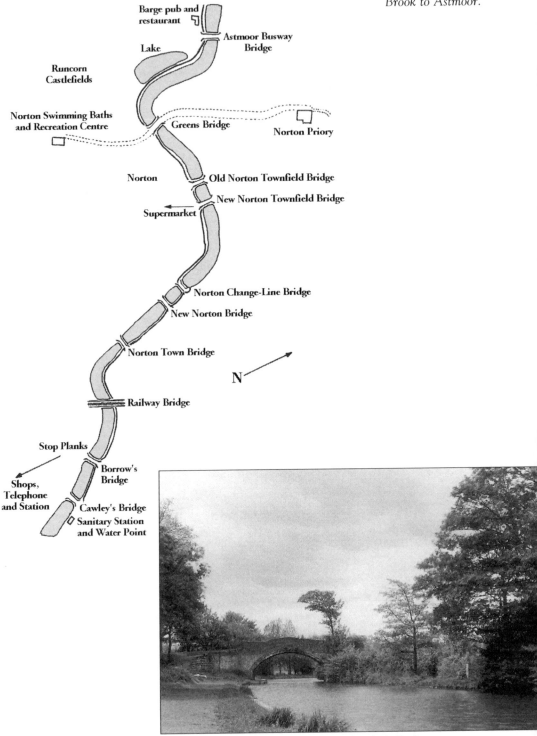

Barge pub and
restaurant

Astmoor Busway
Bridge

Lake

Runcorn
Castlefields

Norton Swimming Baths
and Recreation Centre

Greens Bridge

Norton Priory

Norton

Old Norton Townfield Bridge

New Norton Townfield Bridge

Supermarket

Norton Change-Line Bridge

New Norton Bridge

Norton Town Bridge

N

Railway Bridge

Stop Planks

Borrow's
Bridge

Shops,
Telephone
and Station

Cawley's Bridge

Sanitary Station
and Water Point

Looking north towards Runcorn from Cawley's Bridge.

Norton Town Bridge approaching Norton Priory

In the distance can be seen the toast-rack bridge (now demol-ished) on the approach to Runcorn.

Old Astmoor Change Line Bridge carries the towpath back to its usual side after passing the Norton Priory Estate.

Two views of the canal in a beautiful rural setting at Moore.

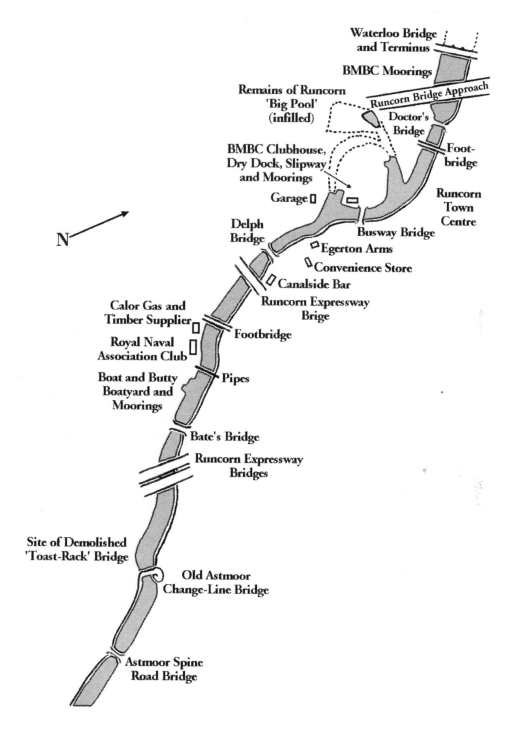

Waterloo Bridge
and Terminus

BMBC Moorings

Runcorn Bridge Approach

Remains of Runcorn
'Big Pool'
(infilled)

Doctor's
Bridge

Foot-
bridge

BMBC Clubhouse,
Dry Dock, Slipway
and Moorings

Garage

Runcorn
Town
Centre

Delph
Bridge

Busway Bridge

Egerton Arms

Convenience Store

Canalside Bar

Runcorn Expressway
Brige

Calor Gas and
Timber Supplier

Royal Naval
Association Club

Footbridge

Boat and Butty
Boatyard and
Moorings

Pipes

Bate's Bridge

Runcorn Expressway
Bridges

Site of Demolished
'Toast-Rack' Bridge

Old Astmoor
Change-Line Bridge

Astmoor Spine
Road Bridge

N

Runcorn Arm. Astmoor to Waterloo Bridge.

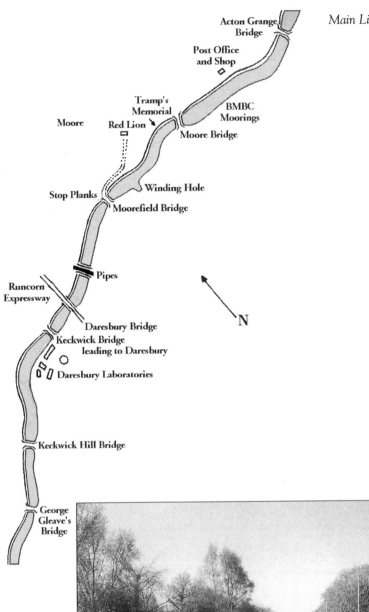

Acton Grange Bridge

Post Office and Shop

Tramp's Memorial

BMBC Moorings

Moore

Red Lion

Moore Bridge

Winding Hole

Stop Planks

Moorefield Bridge

Pipes

Runcorn Expressway

N

Daresbury Bridge

Keckwick Bridge leading to Daresbury

Daresbury Laboratories

Keckwick Hill Bridge

George Gleave's Bridge

Moored craft at Moore.

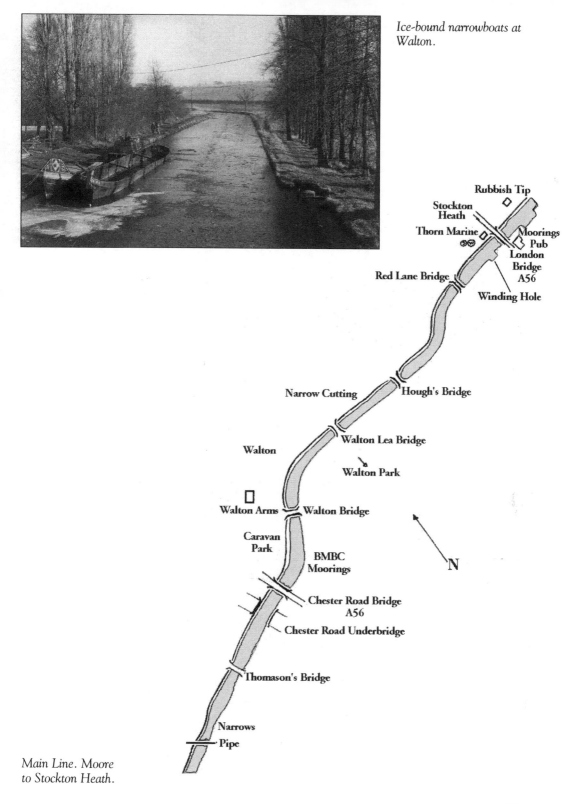

Ice-bound narrowboats at Walton.

Rubbish Tip

Stockton
Heath

Thorn Marine

Moorings
Pub
London
Bridge
A56

Red Lane Bridge

Winding Hole

Narrow Cutting

Hough's Bridge

Walton Lea Bridge

Walton

Walton Park

Walton Arms

Walton Bridge

Caravan
Park

BMBC
Moorings

N

Chester Road Bridge
A56

Chester Road Underbridge

Thomason's Bridge

Narrows

Pipe

*Main Line. Moore
to Stockton Heath.*

The London Bridge pub.

The steps that passengers used to walk down when alighting the passenger fly boats at London Bridge.

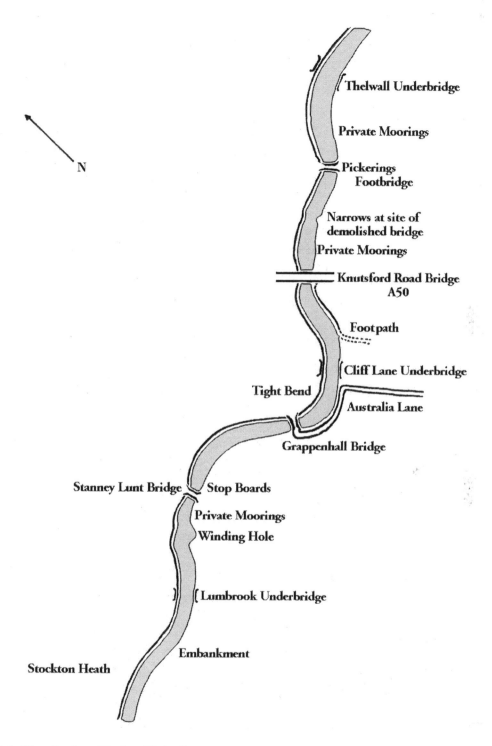

N

Thelwall Underbridge

Private Moorings

Pickerings Footbridge

Narrows at site of demolished bridge

Private Moorings

Knutsford Road Bridge A50

Footpath

Cliff Lane Underbridge

Tight Bend

Australia Lane

Grappenhall Bridge

Stanney Lunt Bridge

Stop Boards

Private Moorings

Winding Hole

Lumbrook Underbridge

Embankment

Stockton Heath

Main Line: Stockton Heath to Thelwall.

The canal at Stockton Heath adjacent to the London Bridge pub.

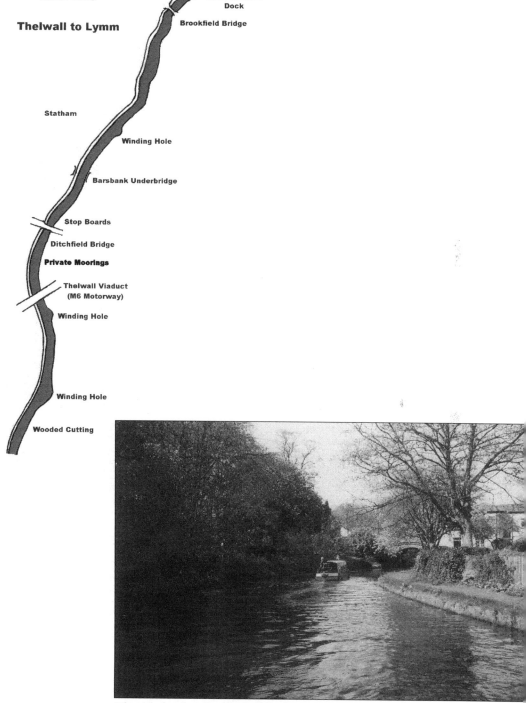

Temporary Moorings
(Both Sides)

Lymm

Main Line

Old Icebreaker
Dock

Thelwall to Lymm

N

Brookfield Bridge

Statham

Winding Hole

Barsbank Underbridge

Stop Boards

Ditchfield Bridge

Private Moorings

**Thelwall Viaduct
(M6 Motorway)**

Winding Hole

Winding Hole

Wooded Cutting

The tight bend at Grappenhall Bridge.

101

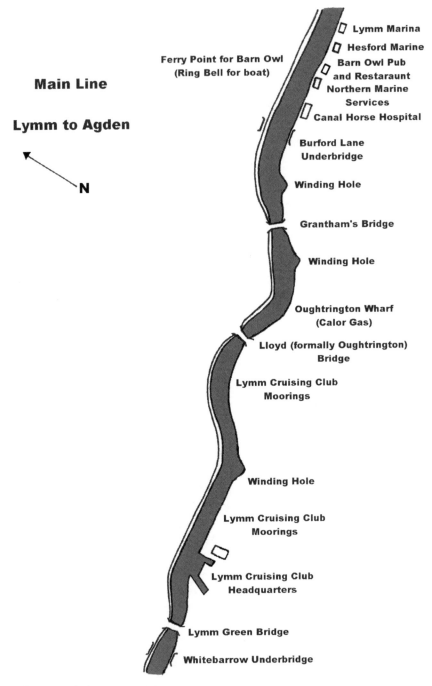

Main Line

Lymm to Agden

N

Lymm Marina

Hesford Marine

Ferry Point for Barn Owl
(Ring Bell for boat)

Barn Owl Pub
and Restaraunt

Northern Marine
Services

Canal Horse Hospital

Burford Lane
Underbridge

Winding Hole

Grantham's Bridge

Winding Hole

Oughtrington Wharf
(Calor Gas)

Lloyd (formally Oughtrington)
Bridge

Lymm Cruising Club
Moorings

Winding Hole

Lymm Cruising Club
Moorings

Lymm Cruising Club
Headquarters

Lymm Green Bridge

Whitebarrow Underbridge

Main Line. Lymm to Agden.

Lymm Bridge.

Lymm Cruising Club's headquarters.

A boat rally at Lymm Cruising Club. The entrance to the Lymm Tunnel is just visible to the left of the white van.

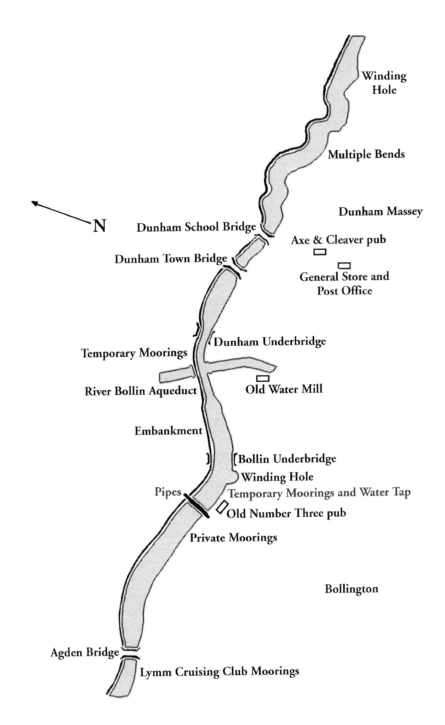

N

Winding Hole

Multiple Bends

Dunham Massey

Dunham School Bridge

Axe & Cleaver pub

Dunham Town Bridge

General Store and Post Office

Dunham Underbridge

Temporary Moorings

River Bollin Aqueduct

Old Water Mill

Embankment

Bollin Underbridge

Winding Hole

Temporary Moorings and Water Tap

Pipes

Old Number Three pub

Private Moorings

Bollington

Agden Bridge

Lymm Cruising Club Moorings

Main Line. Agden to Dunham Massey.

The embankment at Dunham Massey.

Approaching Dunham Massey village.

The Canal winding its way to Altrincham. This photograph was taken prior to the construction of Oldfield Quays Moorings and housing development.

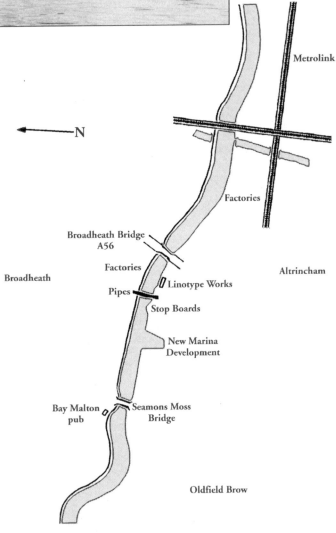

Metrolink

N

Factories

Broadheath Bridge
A56

Broadheath

Factories

Altrincham

Pipes

Linotype Works

Stop Boards

New Marina
Development

Bay Malton
pub

Seamons Moss
Bridge

Oldfield Brow

*Main Line:
Dunham Massey
to Timperley.*

106

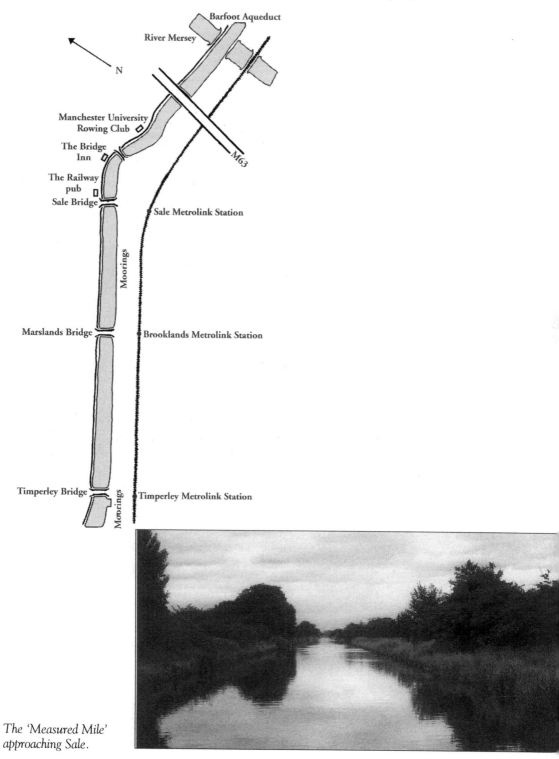

Barfoot Aqueduct

River Mersey

N

Manchester University
Rowing Club

The Bridge
Inn

The Railway
pub

Sale Bridge

M63

Sale Metrolink Station

Moorings

Marslands Bridge

Brooklands Metrolink Station

Timperley Bridge

Timperley Metrolink Station

Moorings

The 'Measured Mile'
approaching Sale.

The Barfoot Aqueduct over the infant River Mersey at Sale. Note the sagging brickwork in the centre of the span.

The River Irwell development approaching Manchester.

Brindley's Circular Weir close to Cornbrook Bridge, Manchester.

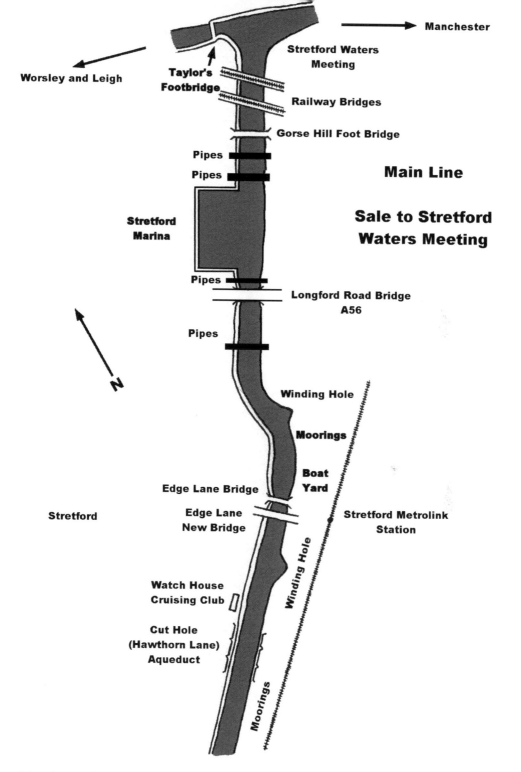

Manchester

Worsley and Leigh

Stretford Waters
Meeting

Taylor's
Footbridge

Railway Bridges

Gorse Hill Foot Bridge

Pipes

Pipes

Main Line

Stretford
Marina

**Sale to Stretford
Waters Meeting**

Pipes

Longford Road Bridge
A56

N

Pipes

Winding Hole

Moorings

Boat
Yard

Edge Lane Bridge

Stretford

Edge Lane
New Bridge

Stretford Metrolink
Station

Winding Hole

Watch House
Cruising Club

Cut Hole
(Hawthorn Lane)
Aqueduct

Moorings

Main Line. Sale to Stretford.

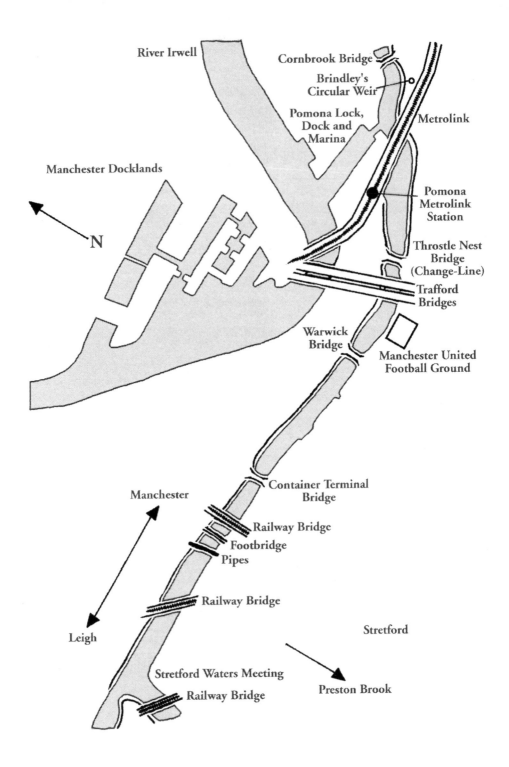

River Irwell

Cornbrook Bridge

Brindley's
Circular Weir

Pomona Lock,
Dock and
Marina

Metrolink

Manchester Docklands

N

Pomona
Metrolink
Station

Throstle Nest
Bridge
(Change-Line)

Trafford
Bridges

Warwick
Bridge

Manchester United
Football Ground

Container Terminal
Bridge

Manchester

Railway Bridge

Footbridge
Pipes

Railway Bridge

Stretford

Leigh

Preston Brook

Stretford Waters Meeting

Railway Bridge

Manchester Branch. Stretford to Manchester Docks.

110

*The now disused
Hulme Lock
approach basin.*

*Views before and
after restoration of
Grocers Warehouse in
Castlefield.*

Barges waiting to be unloaded at Barton Powerstation.

Barton Aqueduct prior to the towpath being removed.

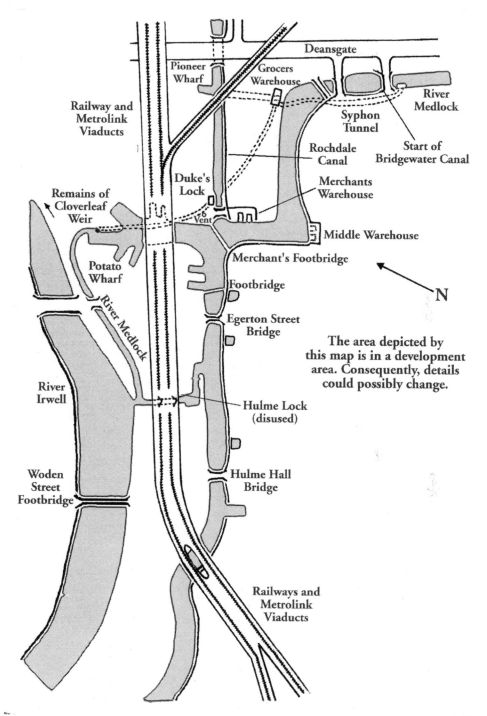

Deansgate

Pioneer Wharf

Grocers Warehouse

River Medlock

Railway and Metrolink Viaducts

Syphon Tunnel

Rochdale Canal

Start of Bridgewater Canal

Duke's Lock

Merchants Warehouse

Remains of Cloverleaf Weir

Vent

Middle Warehouse

Potato Wharf

Merchant's Footbridge

Footbridge

River Medlock

Egerton Street Bridge

The area depicted by this map is in a development area. Consequently, details could possibly change.

N

River Irwell

Hulme Lock (disused)

Woden Street Footbridge

Hulme Hall Bridge

Railways and Metrolink Viaducts

Manchester Branch. Manchester Docks to Castlefield.

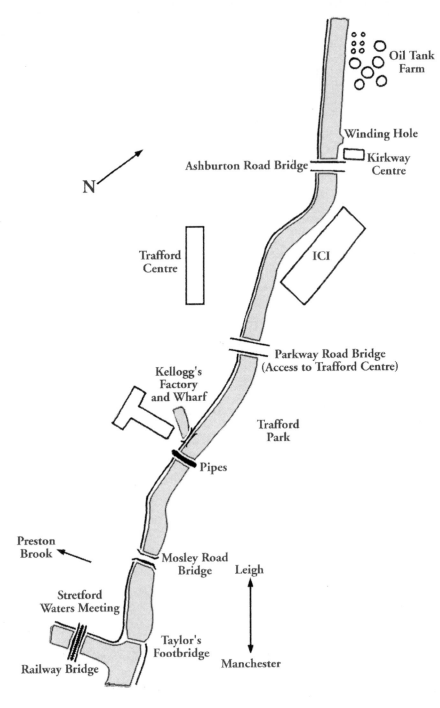

Leigh Branch. Stretford to Barton Aqueduct.

An unusual shot of Barton Swing Aqueduct when it was painted red and white in 1987.

The first railway crosses the first canal. The Liverpool to Manchester Railway viaduct at Monton.

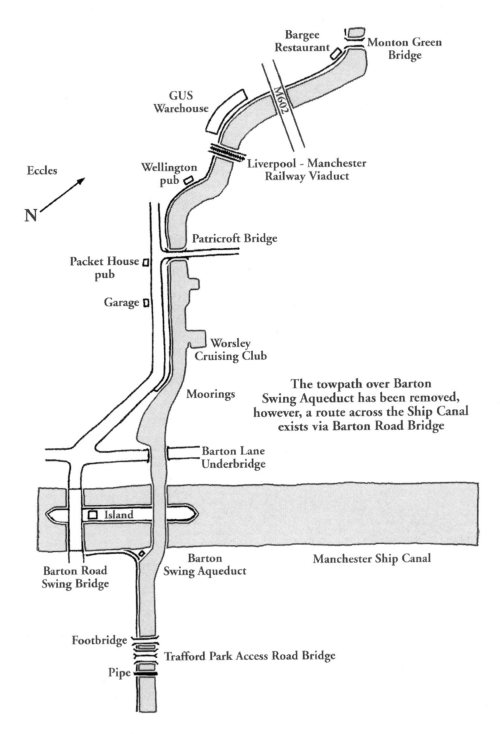

Bargee
Restaurant

Monton Green
Bridge

GUS
Warehouse

M602

Eccles

N

Wellington
pub

Liverpool - Manchester
Railway Viaduct

Patricroft Bridge

Packet House
pub

Garage

Worsley
Cruising Club

Moorings

The towpath over Barton
Swing Aqueduct has been removed,
however, a route across the Ship Canal
exists via Barton Road Bridge

Barton Lane
Underbridge

Island

Manchester Ship Canal

Barton Road
Swing Bridge

Barton
Swing Aqueduct

Footbridge

Trafford Park Access Road Bridge

Pipe

Leigh Branch. Barton Aqueduct to Monton.

116

The Packet House at Worsley Turn.

The new marina development at the former boats' graveyard at Boothstown.

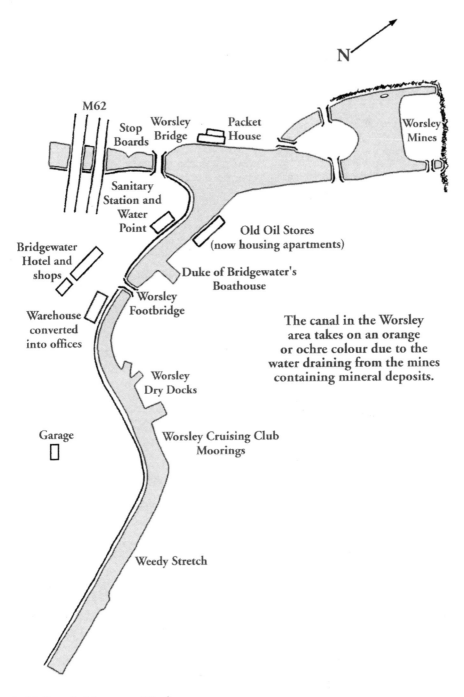

N

M62

Stop Boards

Worsley Bridge

Packet House

Worsley Mines

Sanitary Station and Water Point

Old Oil Stores
(now housing apartments)

Bridgewater Hotel and shops

Duke of Bridgewater's Boathouse

Worsley Footbridge

The canal in the Worsley area takes on an orange or ochre colour due to the water draining from the mines containing mineral deposits.

Warehouse converted into offices

Worsley Dry Docks

Garage

Worsley Cruising Club Moorings

Weedy Stretch

Leigh Branch. Monton to Worsley.

Astley Bridge. Note the increased height of the canal banks to combat the land subsidence in the area.

The disused pit head at Astley Colliery, now home to a mining museum.

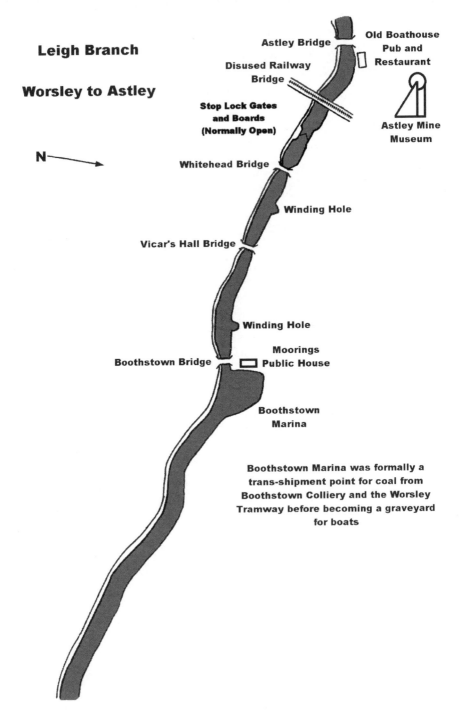

Leigh Branch

Worsley to Astley

N→

Astley Bridge

Old Boathouse
Pub and
Restaurant

Disused Railway
Bridge

Astley Mine
Museum

Stop Lock Gates
and Boards
(Normally Open)

Whitehead Bridge

Winding Hole

Vicar's Hall Bridge

Winding Hole

Moorings
Public House

Boothstown Bridge

Boothstown
Marina

Boothstown Marina was formally a
trans-shipment point for coal from
Boothstown Colliery and the Worsley
Tramway before becoming a graveyard
for boats

Leigh Branch. Worsley to Astley.

Butts Basin at Leigh.

The canal passes through a canyon of factories in Leigh.

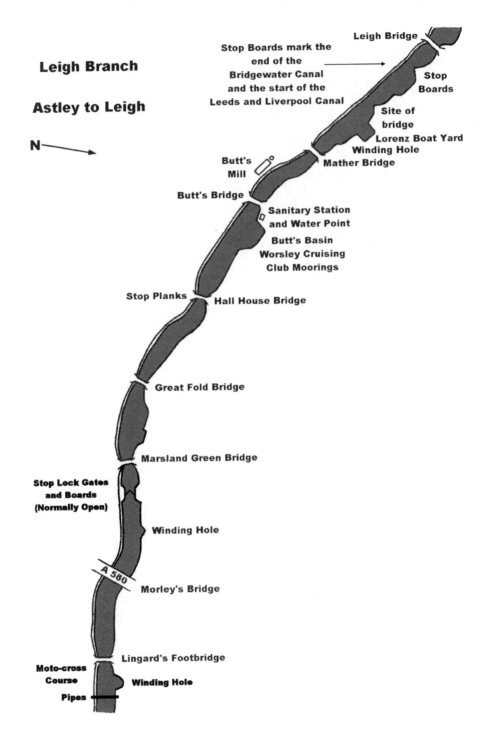

Leigh Branch

Astley to Leigh

N ⟶

Leigh Bridge

Stop Boards mark the end of the Bridgewater Canal and the start of the Leeds and Liverpool Canal ⟶

Stop Boards

Site of bridge

Lorenz Boat Yard

Winding Hole

Mather Bridge

Butt's Mill

Butt's Bridge

Sanitary Station and Water Point

Butt's Basin
Worsley Cruising Club Moorings

Stop Planks

Hall House Bridge

Great Fold Bridge

Marsland Green Bridge

Stop Lock Gates and Boards (Normally Open)

Winding Hole

A 580

Morley's Bridge

Lingard's Footbridge

Moto-cross Course

Winding Hole

Pipes

Leigh Branch. Astley to Leigh.

Four

Navigational & General Information

The Bridgewater Canal is owned, operated and administered by the Bridgewater Department of the Manchester Ship Canal Company, Peel Dome, The Trafford Centre, Manchester, M17 8PL. Telephone number 0161 629 8266; fax number 0161 629 8334. There is also an emergency out-of-hours telephone number which is 0151 327 2212 for Port Security at Eastham Locks.

Maximum Craft Dimensions

Beam	4.2m	(14ft)
Length	21.3m	(70ft)
Air Draught (Headroom)	2.5m	(8ft 6in)
Draught	0.6m	(2ft but deeper craft by arrangement)

Licencing

All craft, including dinghies, canoes and tenders must display a current Bridgewater Canal Licence or British Waterways Licence. If possible, the licence should be displayed at the front of the craft, facing forwards, on the port (left) side of the cabin or in the windscreen. Craft licensed by British Waterways may cruise on the Bridgewater Canal for a period not exceeding seven days. Any period longer than this will require a Bridgewater Canal licence which can be obtained on application for periods of time less than the standard twelve-month licence. Private licence holders are not permitted to trade, hire or carry fare-paying passengers.

Under a reciprocal arrangement with British Waterways, Bridgewater Canal licence holders may cruise on the following stretches of waterway for a period not exceeding seven days, free of charge:

Leeds & Liverpool Canal	Leigh to Burscough and from Leigh to the bottom of Blackburn Locks
Trent & Mersey Canal	Preston Brook Tunnel to Harecastle Tunnel
Shropshire Union Canal	Middlewich to Barbridge Junction
River Weaver	Between Hunts Lock and Saltersford Lock with the usual charges payable for use of the Anderton Lift

The Bridgewater Canal licence also covers the upper reaches of the Manchester Ship Canal from Lowry Footbridge to Hunt's Bank on the River Irwell (the head of navigation) subject to additional conditions available on request from the Manchester Ship Canal Company. All craft must be insured for Third Party Risks and Salvage Costs and have a current Boat Safety Certificate covering current boat construction and use regulations. Craft can only be permanently moored on the canal at approved sites or marinas.

Speed Limit

The maximum speed limit on the canal is 6.5kph (4mph).
If excessive wash is produced, then craft must slow down until an acceptable wash is present. Slow down when passing moored craft and when approaching a bridge, junction, fishermen, small unpowered craft or other navigational hazards.

Every vessel should, at all times, proceed at a safe speed so that the steerer can take proper and effective action to avoid collision and be stopped within a distance appropriate to the prevailing circumstances and conditions.

Mooring

Do not moor
> On a bend.
> Within 23m (25yd) of a bridge.
> On an aqueduct.
> Try to avoid mooring within 9m (10yd) of an angler.
> On the offside of the canal whilst in transit except on an approved twenty-four hour mooring.
> At a water point/sanitary station except when using the facility.
> Opposite a winding hole (used by boats to turn around).
> Where there is an official company mooring prohibition notice.

When mooring
> Tie up bow and stern (front and back).
> Use mooring rings or bollards if available.
> Beware of possible erosions, which may be hidden by grass.
> Beware of possible services laid within the towpath. Look for markers indicating possible dangers, e.g. electricity cables, gas or water.
> Ropes must not cross the towpath as this creates a hazard for pedestrians.
> Mooring adjacent to the towpath is for twenty-four hours only whilst in transit.
> Make sure that the mooring pins are marked with something light or bright to indicate a potential hazard to other towpath users, e.g. upturned plastic bottle or white cloth attached to the top of the pin.

Mooring stakes must be driven into the towpath with care to avoid damaging the canal sidewall. In some places, electricity cables are buried beneath the towpath. These locations are signposted and should be avoided for obvious reasons. The towpath is a Public Right of Way, so mooring ropes must not be placed across it to prevent pedestrians tripping over the ropes. It is forbidden to moor within 21.5m (70ft) of bridges, aqueducts, locks, other obstructions or engineering and maintenance works.

Navigation

Every vessel shall:

At all times, maintain a proper lookout by sight and bearing.

Where possible, keep to the centre of the canal.

Overtake other boats on the port side (left) at normal speed. The boat being overtaken has right of way.

Make sure your way is clear before commencing your manoeuvre. An overtaking manoeuvre must not be made where your visibility is restricted, i.e. on a bend or approaching a bridge or where your wash will disturb anglers or moored craft.

When being overtaken, slow down and travel at a speed sufficient to maintain steerage. Where possible, move over to starboard (right).

Try to avoid turning or manoeuvreing in the vicinity of any anglers.

When approaching narrow bridges, give way to the nearer craft. It has right of way. If you are unsure, give way.

Give way to towing craft. A towing craft has right of way.

All craft shall keep well clear of any dredging or working craft and obey any hazard or speed notices. Working and other wide craft need more room to manoeuvre.

Give an audible notice of approach when necessary. A craft shall give adequate warning by sounding a horn or other suitable device.

Be aware of unpowered craft. Slow down when approaching as many unpowered craft are crewed by small children.

Give due consideration at all times to other users of the canal's facilities and the canal's neighbors.

Navigation at night is not recommended. Every vessel when under way between sunset and sunrise or in conditions of restricted visibility shall, as a minimum requirement, carry a suitable white light visible fore and aft. A navigation light is not a tunnel light. Craft equipped with side navigation lights, e.g. red port (left), green starboard (right) and white masthead (centre), shall also exhibit them.

The total number of persons carried on board the pleasure craft must not exceed the craft's designed carrying capacity. To maintain stability and headroom under bridges whilst underway, crew and passengers must not occupy the roof area of the craft.

Health and Safety

In the event that a member of the crew falls overboard, put engine immediately into neutral. Throw out a line or lifebelt to the person in the water. When safe to do so, advance the boat slowly towards the bank to make a rescue or move the craft near to the person in the water where he or she can be pulled back on board. Remember that the person in the water may be at greatest risk from the propeller. Never put the engine in gear to turn the propeller until it is safe to do so.

Waterborne Diseases

Certain waterborne diseases associated with vermin may be present in canal water. Users of the canal and adjacent land should pay due respect to cleanliness, particularly where open wounds, e.g. cuts, abrasions, etc., are involved and should always wash their hands before eating. Anyone accidentally falling into the canal should seek medical attention as soon as possible as a precautionary measure.

Children and Pets

Children and pets should be under supervision at all times.
Arms and legs in particular should be kept clear of potential crush hazards, e.g. canal banks, bridges and other craft.
Children and all non-swimmers should wear lifejackets at all times.
Obey any signs ordering dogs to be kept on a lead.
Abide by any emergency measures controlling animals that are in force.

Swimming

To swim or wade in canals is dangerous and is not permitted. Swimming may result in death. The dangers are:
    Poor water quality.
    Depth of water.
    Coldness of water.
    Underwater obstructions.
    Presence of silt.
    Swimmers may become entangled in rubbish or weed.
    Swimmers may be hit by a passing boat.
    The water depth is up to 2m. Children should be supervised at all times.
    To swim in the Bridgewater Canal is an offence under the Bridgewater Canal By-Laws 1961.

Ice

Never walk on the canal (or any other deep water for that matter) when covered with ice. The thickness of the ice cannot be measured and is unpredictable.

Pollution and Waste Disposal

Always take your rubbish home or dispose of it into skips and bins provided by the Company. Sewage, oil or oil derivatives, toilet waste or any other pollutants must not be discharged into the canal at any time.
In the event of a pollution incident please contact the Company or Environmental Agency (0800 807 060) at the earliest opportunity, giving full details of the incident.

Dog Fouling

The fouling of the towpath by dogs is not permitted. Dog owners are asked to show consideration for other users of the towpath by cleaning up after their pet.

General Information

Angling

A current Angling Licence for the stretch of canal being fished must be obtained.

Exercise due consideration at all times for other users of the canal, its facilities and the canal's neighbours.

The towpath is a Public Right of Way in many places. Take care that your fishing equipment is not blocking the towpath.

Before setting up to fish always check for powerlines and never fish within 30m (100ft) of them. Look out for pedestrians and others when casting or drawing back.

Do not leave your rod or pole across the towpath when not in use.

In the interests of safety, anglers are advised not to raise the pole or rod over a passing boat. It is much better to either pull back or place it parallel to the canal wall.

Do not fish within 9m (30ft) of a bridge.

Take your litter and unused bait home. Discarding bait or food will spoil other people's fishing and may attract vermin.

To avoid disturbance by boats turning, do not fish in winding holes (wide areas of water used by boats to turn around).

Anglers should not shout instructions to passing craft. Craft will normally keep to the centre of the canal.

Do not fish within 5m (16ft) of a moored boat.

When fishing opposite a moored boat, take care not to allow ground bait etc. to come into contact with the boat.

The fishing rights on the Bridgewater Canal are leased to a number of different fishing clubs, information on which is available from the Manchester Ship Canal Company.

Use of Towpaths

The following points are to be observed at all times when using the towpaths:

Motorcycling on the towpaths is not permitted.

Cycling on the towpaths is not permitted.

Horse riding on the towpaths is not permitted.

Due consideration is to be given at all times to other users of the facilities and the canal's neighbours.

Footwear should be worn to suit the towpath and weather conditions

In certain areas the towpath may not be suitable for prams or wheelchair use.

The towing path is unlit and due to the close proximity of deep water care must be taken in poor light conditions.

Bank erosion is caused by water action washing against the walls of the canal. Sometimes, holes can appear at the back of the coping stones and these holes may be hidden by grass.

Care must be taken when alighting from boats and when using the towpath.

Historic features, such as mooring rings, posts, etc., may obstruct the towpath.

Mooring ropes and fishing tackle may also present a hazard for pedestrians.

Barriers have been erected across the towpath at various locations to prevent access by motorcycles.

Please take your uneaten food and litter home to dispose of it properly. Carelessly discarded food may attract vermin.

If an angler is baiting his line, the pole or rod may be across the towpath. Please wait a moment or two for the pole or rod to be removed.

Footnote

Wherever I go, my interest in canals seems to follow me. If I am travelling by road or train and see the black and white of a lock gate balance beam or the unique profile of a humpbacked bridge, I try to identify the canal or waterway from a mental map. I also get the urge to stop for a look around and to take photographs (a good photographer always carries a camera) and take notes.

Recently, whilst on a holiday to Majorca, I unexpectedly came across humpbacked bridges and a canal system connecting lagoons and a marina to the sea, located in a town called Alcudia. Not only did the town possess a small canal network but I also discovered a British *Norman* 24ft canal cruiser tucked away in a residential area adjacent to the canal. It was on a trailer, scruffy and had no windows but was readily identifiable for what it was: a British, narrow beam, fibreglass canal cruiser. I wonder if the owner knows why it is only 6ft 10in beam? No matter where I travel, associations to my beloved canals and inland waterways rarely seem very far away.

One thing that I find very frustrating when writing this (or any other of my titles for that matter) is that just after I have produced the 'final draft' of the text, I discover additional information, remember something that I should have added or something happens to be worthy of inclusion in the book. I think that any book of this type cannot be considered finished; today's happenings are tomorrow's history. With this in mind, I look forward to writing the 'second edition' with relish and wonder what the future holds in store for the Duke's Cut, alias the Bridgewater Canal, Britain's first true canal.

About the Author

Cyril J. Wood has had an active interest in canals and inland waterways since a child when, in 1960, his parents hired a cabin cruiser from Dean's Pleasure Cruisers at Christleton on the Shropshire Union Canal.

He is a qualified photographer and college lecturer in photography. As well as being a prolific photographer and producing audio/visual presentations in his spare time, to date he has had two books published: *The Duke's Cut – The Bridgewater Canal* and *The Big Ditch – The Manchester Ship Canal*, both published originally by Tempus (now the History Press). In addition to these two works he has also written photography text books (Theory, History and Practical), *Wyre Heal – A Local History of the Wirral Peninsula, From Navy League to Wirral Met – The History of Wirral Metropolitan College and its Campuses* (in eBook format), plus a series of children's novels and articles for many magazines on subjects ranging from photography, ciné & video, audio/visual and Wirral's local history to canals and inland waterways.

Cyril has also had involvement with other forms of media as well as photography and the written word. He is Radio Merseyside's canals and inland waterways correspondent and has been involved in many programmes for the station. He was also contacted by researchers from BBC's television series *What The Industrial Revolution Did For Us* regarding locations for the episode that concentrated on canals and *The Duke's Cut* was extensively used for research during the programme. He was also interviewed by John Craven on the BBC's *Countryfile* programme when they produced a feature on the Manchester Ship Canal. Most of the archive photographs used in the programme were from *The Big Ditch* as well as facts and other information contained within the book.

Lastly and most importantly (to me anyway...) Cyril introduced me to canal cruising for which I shall be eternally grateful.

John White
(A friend of Cyril's ... not the boatbuilder)